Nonsmooth Optimization in Honor of the 60th Birthday of Adil M. Bagirov

Nonsmooth Optimization in Honor of the 60th Birthday of Adil M. Bagirov

Editors

Napsu Karmitsa
Sona Taheri

MDPI • Basel • Beijing • Wuhan • Barcelona • Belgrade • Manchester • Tokyo • Cluj • Tianjin

Editors
Napsu Karmitsa Sona Taheri
University of Turku RMIT University
Finland Australia

Editorial Office
MDPI
St. Alban-Anlage 66
4052 Basel, Switzerland

This is a reprint of articles from the Special Issue published online in the open access journal *Algorithms* (ISSN 1999-4893) (available at: https://www.mdpi.com/si/algorithms/Nonsmooth_Optimization).

For citation purposes, cite each article independently as indicated on the article page online and as indicated below:

LastName, A.A.; LastName, B.B.; LastName, C.C. Article Title. *Journal Name* **Year**, *Volume Number*, Page Range.

ISBN 978-3-03943-835-8 (Hbk)
ISBN 978-3-03943-836-5 (PDF)

© 2020 by the authors. Articles in this book are Open Access and distributed under the Creative Commons Attribution (CC BY) license, which allows users to download, copy and build upon published articles, as long as the author and publisher are properly credited, which ensures maximum dissemination and a wider impact of our publications.

The book as a whole is distributed by MDPI under the terms and conditions of the Creative Commons license CC BY-NC-ND.

Contents

About the Editors vii

Preface to "Nonsmooth Optimization in Honor of the 60th Birthday of Adil M. Bagirov" . . . ix

Napsu Karmitsa and Sona Taheri
Special Issue "Nonsmooth Optimization in Honor of the 60th Birthday of Adil M. Bagirov":
Foreword by Guest Editors
Reprinted from: *Algorithms* **2020**, *13*, 282, doi:10.3390/a13110282 1

Bruno Colonetti, Erlon Cristian Finardi and Welington de Oliveira
A Mixed-Integer and Asynchronous Level Decomposition with Application to the Stochastic
Hydrothermal Unit-Commitment Problem
Reprinted from: *Algorithms* **2020**, *13*, 235, doi:10.3390/a13090235 5

Marek J. Śmietański
On a Nonsmooth Gauss–Newton Algorithms for Solving Nonlinear Complementarity
Problems
Reprinted from: *Algorithms* **2020**, *13*, , doi:10.3390/a13080190 21

Andreas Griewank and Andrea Walther
Polyhedral DC Decomposition and DCA
Optimization of Piecewise Linear Functions
Reprinted from: *Algorithms* **2020**, *13*, 166, doi:10.3390/a13070166 33

Angel A. Juan, Canan G. Corlu, Rafael D. Tordecilla, Rocio de la Torre, and Albert Ferrer
On the Use of Biased-Randomized Algorithms for Solving Non-Smooth Optimization Problems
Reprinted from: *Algorithms* **2020**, *13*, 8, doi:10.3390/a13010008 59

Outi Montonen, Timo Ranta and Marko M. Mäkelä
Planning the Schedule for the Disposal of the Spent Nuclear Fuel with Interactive
Multiobjective Optimization
Reprinted from: *Algorithms* **2019**, *12*, 252, doi:10.3390/a12120252 73

Annabella Astorino, Antonio Fuduli, Giovanni Giallombardo and Giovanna Miglionico
SVM-Based Multiple Instance Classification via DC Optimization
Reprinted from: *Algorithms* **2019**, *12*, 249, doi:10.3390/a12120249 93

About the Editors

Napsu Karmitsa has been a Docent (Adjunct Professor) of Applied Mathematics at the Department of Mathematics and Statistics at the University of Turku, Finland, since 2011. She obtained her MSc degree in Organic Chemistry in 1998 and Ph.D. degree in Scientific Computing in 2004, both from the University of Jyväskylä, Finland. At the moment, she holds a position of Academy Research Fellow granted by the Academy of Finland. Her research is focused on nonsmooth optimization and analysis. Special emphasis is given to nonconvex, global, and large-scale cases with applications in machine learning and data analysis. Dr. Karmitsa has published two books on nonsmooth optimization and one book on clustering as well as several journal papers and research reports. In addition, her webpage http://napsu.karmitsa.fi is one of the leading sources of nonsmooth optimization solvers available.

Sona Taheri received her Ph.D. in Information Technology from Federation University Australia in 2012. Currently, she holds a lecturer position at RMIT University, Australia. She has conducted research in the area of optimization, particularly nonsmooth, nonconvex, and DC optimization; data mining and machine learning, particularly cluster analysis and regression in large data sets; and cybersecurity, mainly in alert analysis and malicious multisource data sets. The results of her research have been published as a book, an edited book, book chapters, and journal and conference articles.

Preface to "Nonsmooth Optimization in Honor of the 60th Birthday of Adil M. Bagirov"

Nonsmooth optimization (NSO) refers to the general problem of minimizing (or maximizing) functions that are typically not differentiable at their minimizers (maximizers). Such functions can be found in various applications such as image denoising; optimal shape design; computational chemistry and physics; water management; cybersecurity; machine learning; and data mining, including cluster analysis, classification, and regression. As the classical optimization theory presumes differentiability of the functions to be optimized, it cannot be directly applied, nor can the methods introduced for smooth problems.

The aim of this book was to gather the most recent developments in NSO techniques and applications. The book opens with the foreword by the Guest Editors and then presents six articles in the area of NSO and its applications.

The Guest Editors are grateful to Professor Adil Bagirov, with whom they have had the privilege of conducting research in the area of NSO and its applications, and wish him all the best in his career and personal life.

The Guest Editors would like to thank all the authors for their contributions to this book, all reviewers who provided constructive comments, and the editorial staff of the MDPI journal *Algorithms* for their support.

Napsu Karmitsa, Sona Taheri
Editors

Editorial

Special Issue "Nonsmooth Optimization in Honor of the 60th Birthday of Adil M. Bagirov": Foreword by Guest Editors

Napsu Karmitsa [1],* and Sona Taheri [2]

[1] Department of Mathematics and Statistics, University of Turku, FI-20014 Turku, Finland
[2] School of Science, RMIT University, Melbourne, VIC 3000, Australia; sona.taheri@rmit.edu.au
* Correspondence: napsu@karmitsa.fi

Received: 4 November 2020; Accepted: 4 November 2020; Published: 7 November 2020

Abstract: Nonsmooth optimization refers to the general problem of minimizing (or maximizing) functions that have discontinuous gradients. This Special Issue contains six research articles that collect together the most recent techniques and applications in the area of nonsmooth optimization. These include novel techniques utilizing some decomposable structures in nonsmooth problems—for instance, the difference-of-convex (DC) structure—and interesting important practical problems, like multiple instance learning, hydrothermal unit-commitment problem, and scheduling the disposal of nuclear waste.

1. Introduction

In this special issue, we take the opportunity to acknowledge the outstanding contributions of Professor Adil Bagirov (Figure 1) to nonsmooth optimization (NSO) in both theoretical foundations and its practical aspects during his 35 year long research career. This Special Issue collects together the most recent techniques and applications in the area of NSO. It contains six excellent research papers by well-known mathematicians. Some of the authors have at some point collaborated with Adil Bagirov, and all of them would like to show their respect to him and his work.

Figure 1. Professor Adil Bagirov.

Adil Bagirov received a master's degree in Applied Mathematics from Baku State University, Azerbaijan in 1983, and the Candidate of Sciences degree in Mathematical Cybernetics from the

Institute of Cybernetics of Azerbaijan National Academy of Sciences in 1989. Then he worked at the Space Research Institute (Baku, Azerbaijan), Baku State University (Baku, Azerbaijan) and Joint Institute for Nuclear Research (Moscow, Russia) until 1998.

Bagirov has been joined with Federation University Australia since 1999. He completed his PhD in Optimization under the supervision of Professor Alexander Rubinov at Federation University Australia (formerly the University of Ballarat) in 2002. Currently, he holds the full Professor position at this university. Professor Bagirov has contributed exceptionally to NSO and its applications to real-life problems. These contributions include writing books on NSO [1] and its applications in clustering [2], an edited book on NSO methods [3] and more than 170 journal papers, book chapters and papers in conference proceedings in the area of NSO and its applications (see, e.g., [4–12]). He has also supervised more than 28 PhD students.

Professor Bagirov has been successful in securing five grants from the Australian Research Council's Discovery and Linkage schemes to conduct research in nonsmooth and global optimization and their applications. He was awarded the Australian Research Council Postdoctoral Fellowship and the Australian Research Council Research Fellowship. In addition, he is EUROPT Fellow 2009.

The Guest Editors are grateful to Professor Adil Bagirov, with whom they have had the privilege to do research in the area of NSO and its real-life applications. On behalf of the journal, the Guest Editors wish him all the best in his career and personal life.

2. Nonsmooth Optimization

NSO refers to the general problem of minimizing (or maximizing) functions that have discontinuous gradients. These types of functions arise in many applied fields, for instance, in image deionising, optimal shape design, computational chemistry and physics, water management, cyber security, machine learning, and data mining including cluster analysis, classification and regression. In most of these applications, the number of decision variables is very large, and their NSO formulations allow us to reduce these numbers significantly. Thus, the application of NSO approaches facilitates the design of efficient algorithms for their solutions, the more realistic modeling of various real-world problems, the robust formulation of a system, and even the solving of difficult smooth (continuously differentiable) problems that require reducing the problem's size or simplifying its structure. These are some of the main reasons for the increased attraction to nonsmooth analysis and optimization during the past few years. This Special Issue collects some of the most recent methods in NSO and its applications. These include novel techniques for solving NSO problems by utilizing, for instance, the decomposable (difference of convex (DC)) structure of the objective, the nonsmooth Gauss-Newton algorithm, the biased-randomized algorithm, and also interesting practical problems such as the multiple instance learning, the hydrothermal unit-commitment problem, and scheduling the disposal of nuclear waste.

In the first article, "A Mixed-Integer and Asynchronous Level Decomposition with Application to the Stochastic Hydrothermal Unit-Commitment Problem" by Bruno Colonetti, Erlon Cristian Finardi and Welington de Oliveira [13], the authors develop an efficient algorithm for solving uncertain unit-commitment (UC) problems. The efficiency of the algorithm is based on the novel asynchronous level decomposition of the UC problem and the parallelization of the algorithm.

In the second article "On a Nonsmooth Gauss-Newton Algorithm for Solving Nonlinear Complementarity Problems" by Marek J. Śmietański [14], the author proposes a new nonsmooth version of the generalized damped Gauss-Newton method for solving nonlinear complementarity problems. In the proposed algorithm, the Bouligand differential plays the role of the derivative. The author presents two types of algorithms (usual and inexact), which have superlinear and global convergence for semismooth cases.

In the article "Polyhedral DC Decomposition and DCA Optimization of Piecewise Linear Functions" by Andreas Griewank and Andrea Walther [15], the abs-linear representation of the piecewise linear functions is extended, yielding their DC decomposition as well as a pair of generalized gradients that can be computed using the reverse mode of algorithmic differentiation. The DC decomposition and two subgradients are used to drive DCA algorithms where the (convex) inner problem can be solved in a finite many iterations and the gradients of the concave part can be updated using a reflection technique.

The fourth article, "On the Use of Biased-Randomized Algorithms for Solving Non-Smooth Optimization Problems" by Angel Alejandro Juan, Canan Gunes Corlu, Rafael David Tordecilla, Rocio de la Torre and Albert Ferrer [16], introduces the use of biased-randomized algorithms as an effective methodology to cope with NP-hard and NSO problems in many practical applications, in particular, those including so called soft constraints. Biased-randomized algorithms extend constructive heuristics by introducing a nonuniform randomization pattern into them. Thus, they can be used to explore promising areas of the solution space without the limitations of gradient-based approaches that assume the existence of the smooth objective.

In the fifth article, "Planning the Schedule for the Disposal of the Spent Nuclear Fuel with Interactive Multiobjective Optimization" by Outi Montonen, Timo Ranta and Marko M. Mäkelä [17], the very important problem of the scheduling of nuclear waste disposal is modelled as a multiobjective mixed-integer nonlinear NSO problem with the minimization of nine objectives. A novel method using the two-slope parameterized achievement scalarizing functions is introduced for solving this problem, and a case study adapting the disposal in Finland is given.

Finally, the article "SVM-Based Multiple Instance Classification via DC Optimization" by Annabella Astorino, Antonio Fuduli, Giovanni Giallombardo and Giovanna Miglionico considers the binary classification of the multiple instance learning problem [18]. The problem is formulated as a nonconvex unconstrained NSO problem with a DC objective function, and an appropriate nonsmooth DC algorithm is used to solve this problem.

The Guest Editors would like to thank all the authors for their contributions in this Special Issue. They would also like to thank all the reviewers for their timely and insightful comments on the submitted articles as well as the editorial staff of the MDPI Journal Algorithms for their assistance in managing the review process in a prompt manner.

Funding: This work was financially supported by Academy of Finland grant #289500.

Conflicts of Interest: The authors declare no conflict of interest.

References

1. Bagirov, A.M.; Karmitsa, N.; Mäkelä, M.M. *Introduction to Nonsmooth Optimization: Theory, Practice and Software*; Springer: Cham, Switzerland, 2014.
2. Bagirov, A.M.; Karmitsa, N.; Taheri, S. *Partitional Clustering via Nonsmooth Optimization: Clustering via Optimization*; Springe: Cham, Switzerland, 2020.
3. Bagirov, A.M.; Gaudioso, M.; Karmitsa, N.; Mäkelä M.M.; Taheri, S. *Numerical Nonsmooth Optimization: State of the Art Algorithms*; Springer: Cham, Switzerland, 2020.
4. Bagirov, A.M. Modified global k-means algorithm for sum-of-squares clustering problems. *Pattern Recognit.* **2008**, *41*, 3192–3199. [CrossRef]
5. Bagirov, A.M.; Clausen, C.; Kohler, M. Estimation of a regression function by maxima of minima of linear functions. *IEEE Trans. Inf. Theory* **2009**, *55*, 833–845. [CrossRef]
6. Bagirov, A.M.; Karasozen, B.; Sezer, M. Discrete gradient method: Derivative-free method for nonsmooth optimization. *J. Optim. Theory Appl.* **2008**, *137*, 317–334. [CrossRef]
7. Bagirov, A.M.; Mahmood, A.; Barton, A. Prediction of monthly rainfall in Victoria, Australia: Clusterwise linear regression approach. *Atmos. Res.* **2017**, *188*, 20–29. [CrossRef]
8. Bagirov, A.M.; Rubinov, A.M.; Zhang, J. A multidimensional descent method for global optimization. *Optimization* **2009**, *58*, 611–625. [CrossRef]
9. Bagirov, A.M.; Taheri, S.; Karmitsa, N.; Joki K.; Mäkelä, M.M. Aggregate subgradient method for nonsmooth DC optimization. *Optim. Lett.* **2020**, in press. [CrossRef]

10. Bagirov, A.M.; Taheri, S.; Ugon, J. Nonsmooth DC programming approach to the minimum sum-of-squares clustering problems. *Pattern Recognit.* **2016**, *53*, 12–24. [CrossRef]
11. Joki, K.; Bagirov, A.M.; Karmitsa, N.; Mäkelä, M.M.; Taheri, S. Double bundle method for finding Clarke stationary points in nonsmooth DC programming. *SIAM J. Optim.* **2018**, *28*, 1892–1919. [CrossRef]
12. Karmitsa, N.; Bagirov, A.M.; Taheri, S. New diagonal bundle method for clustering problems in large data sets. *Eur. J. Oper. Res.* **2017**, *263*, 367–379. [CrossRef]
13. Colonetti, B.; Finardi, E.C.; de Oliveira, W. A Mixed-Integer and Asynchronous Level Decomposition with Application to the Stochastic Hydrothermal Unit-Commitment Problem. *Algorithms* **2020**, *13*, 235. [CrossRef]
14. Śmietański, M.J. On a Nonsmooth Gauss–Newton Algorithms for Solving Nonlinear Complementarity Problems. *Algorithms* **2020**, *13*, 190. [CrossRef]
15. Griewank, A.; Walther, A. Polyhedral DC Decomposition and DCA Optimization of Piecewise Linear Functions. *Algorithms* **2020**, *13*, 166. [CrossRef]
16. Juan, A.A.; Corlu, C.G.; Tordecilla, R.D.; de la Torre, R.; Ferrer, A. On the Use of Biased-Randomized Algorithms for Solving Non-Smooth Optimization Problems. *Algorithms* **2020**, *13*, 8. [CrossRef]
17. Montonen, O.; Ranta, T.; Mäkelä, M.M. Planning the Schedule for the Disposal of the Spent Nuclear Fuel with Interactive Multiobjective Optimization. *Algorithms* **2019**, *12*, 252. [CrossRef]
18. Astorino, A.; Fuduli, A.; Giallombardo, G.; Miglionico, G. SVM-Based Multiple Instance Classification via DC Optimization. *Algorithms* **2019**, *12*, 249. [CrossRef]

Publisher's Note: MDPI stays neutral with regard to jurisdictional claims in published maps and institutional affiliations.

© 2020 by the authors. Licensee MDPI, Basel, Switzerland. This article is an open access article distributed under the terms and conditions of the Creative Commons Attribution (CC BY) license (http://creativecommons.org/licenses/by/4.0/).

Article

A Mixed-Integer and Asynchronous Level Decomposition with Application to the Stochastic Hydrothermal Unit-Commitment Problem

Bruno Colonetti [1],*, Erlon Cristian Finardi [1,2] and Welington de Oliveira [3],*

1. Department of Electrical and Electronic Engineering, Federal University of Santa Catarina, Florianópolis 88040-900, Brazil; erlon.finardi@ufsc.br
2. INESC P&D Brasil, Bairro Gonzaga 11055-300, Brazil
3. MINES ParisTech, CMA—Centre de Mathématiques Appliquées, PSL—Research University, Sophia Antipolis, 75006 Paris, France
* Correspondence: colonetti.bruno@posgrad.ufsc.br (B.C.); welington.oliveira@mines-paristech.fr (W.d.O.)

Received: 3 August 2020; Accepted: 14 September 2020; Published: 18 September 2020

Abstract: Independent System Operators (ISOs) worldwide face the ever-increasing challenge of coping with uncertainties, which requires sophisticated algorithms for solving unit-commitment (UC) problems of increasing complexity in less-and-less time. Hence, decomposition methods are appealing options to produce easier-to-handle problems that can hopefully return good solutions at reasonable times. When applied to two-stage stochastic models, decomposition often yields subproblems that are embarrassingly parallel. Synchronous parallel-computing techniques are applied to the decomposable subproblem and frequently result in considerable time savings. However, due to the inherent run-time differences amongst the subproblem's optimization models, unequal equipment, and communication overheads, synchronous approaches may underuse the computing resources. Consequently, asynchronous computing constitutes a natural enhancement to existing methods. In this work, we propose a novel extension of the asynchronous level decomposition to solve stochastic hydrothermal UC problems with mixed-integer variables in the first stage. In addition, we combine this novel method with an efficient task allocation to yield an innovative algorithm that far outperforms the current state-of-the-art. We provide convergence analysis of our proposal and assess its computational performance on a testbed consisting of 54 problems from a 46-bus system. Results show that our asynchronous algorithm outperforms its synchronous counterpart in terms of wall-clock computing time in 40% of the problems, providing time savings averaging about 45%, while also reducing the standard deviation of running times over the testbed in the order of 25%.

Keywords: stochastic programming; stochastic hydrothermal UC problem; parallel computing; asynchronous computing; level decomposition

1. Introduction

The unit-commitment (UC) problem aims at determining the optimal scheduling of generating units to minimize costs or maximize revenues while satisfying local and system-wide constraints [1]. In its deterministic form, UC still poses a challenge to operators and researchers due to the large sizes of the systems and the increasing modeling details necessary to represent the system operation. For instance, in the Brazilian case, the current practice is to set a limit of 2 h for the solution of the deterministic UC [2], while the Midcontinent Independent System Operator (MISO) sets a time limit of 20 min for its UC [3]. (Note that the Brazilian system and the MISO are different from a physical, as well as from a market-based, viewpoint, but the problems being solved in these two cases share the same classical concept of the UC.) Nonetheless, the growing presence of intermittent generation has added yet more difficulty to the problem, giving rise to what is called uncertain UC [4]. The latter is considerably harder to solve than its deterministic counterpart, and one of the reasons for its lack of adoption in the industry is precisely its computational burden: Large-scale uncertain UC takes a prohibitively long time to be solved. In this context, efficient solution methods for the uncertain UC that can take full advantage of the computational resources at hand are both desirable and necessary to help system operators cope with uncertain resources.

In particular, to model the uncertainty arising from renewable sources, one of two approaches is generally employed: robust optimization or stochastic programming [4]. The latter is by far the most employed, both in its chance-constrained and recourse variants. In stochastic programs with recourse, uncertainty is, in general, represented by finite-many scenarios, and the problem is formulated either in a two-stage or multistage setting. In two-stage stochastic problems, the first-stage variables must be decided before uncertainty is revealed. Once the uncertain information becomes known, recourse actions are taken to best accommodate the first-stage decisions [5]. In stochastic hydrothermal unit-commitment (SHTUC) problems, the sources of uncertainties are related to renewable resources, spot prices, load, and equipment availability [1,4].

The commitment decisions are usually modeled as first-stage variables, while dispatch decisions are the recourse actions (second-stage variables). Given the mixed-integer nature of commitment decisions, SHTUC problems in a two-stage formulation give rise to large-scale mixed-integer optimization models whose numerical solution by off-the-shelf solvers is often prohibitive due to time requirements or limited computing resources. Consequently, decomposition techniques must come into play [1,4,6,7]. Benders decomposition (BD) and Lagrangian relaxation (LR) are the most used techniques to handle SHTUC problems. While the BD deals with the primal problem [8], LR is a dual procedure employed to compute the best lower bound for the SHTUC problem [7,9]. Primal-recovery heuristics are employed to compute primal-feasible points, which are not, in general, optimal solutions. This is the main shortcoming of LR-based techniques.

Decomposition techniques yield models that are amenable for parallelization [5]. A common strategy for solving problems simultaneously is to use a master/worker framework with pre-specified synchronization points [10], which we call synchronous computing (SYN). In this framework, the master chooses new iterates and sends them to workers, who, in turn, are responsible for solving one or more subproblems. Examples of SYN implementations for UC are given in [11–14]. An aspect of SYN is that, at predetermined points of the algorithm, the master must wait for all workers to respond to resume the iterative process: the synchronization points. However, the times for workers to finish their respective tasks might vary significantly. This results in idle times, both for the master and for workers who respond quickly [10]. One way to reduce idle times is to use asynchronous computing (ASYN).

In contrast to SYN, in ASYN, there are no synchronization points, so the master and workers do not need to wait until all workers respond to continue their operations. Thus, in an iterative process, e.g., in BD, the master would compute the next iterate based on information of possibly only a proper, but nonempty, subset of the workers. Based on this possibly incomplete information, the master sends a new iterate to available workers, while slower workers are still carrying their tasks on an outdated iterate. Because in ASYN iterates might not be evaluated by all workers, the evaluation of the objective function (yielding bounds on the optimal values) is precluded. Hence, a fundamental step in ASYN is the (scarce) coordination of workers to produce valid bounds.

ASYN implementations have been proposed in the UC literature mainly to solve the dual problems (issued by LRs) via either subgradient algorithms or cutting-plane-based methods [15–17]. In References [15,16], a queue of iterates is created and its elements are gradually sent to the workers. Auxiliary lists keep track of the evaluation status of each worker with respect to the elements in the queue. Once an element has been evaluated by all workers, a valid bound to the original problem is available. The authors of Reference [15] demonstrate that their algorithm converges to a dual global solution regardless of the iterate-selection policy used to choose the iterates from the queue—first-in-first-out or last-in-first-out. In References [17], the authors keep a list of all the iterates to compute valid bounds. In addition to solving the dual problem asynchronously, Reference [17] also conducts the primal recovery asynchronously. While References [15,16] employ a convex trust-region bundle method, Reference [17] implements an incremental subgradient method. Asynchronous implementations of BD for convex problems can be found in References [18–20]. In Reference [18], the dual dynamic-programming algorithm is handled asynchronously in a hydrothermal scheduling problem. In Reference [19], the stochastic dual dynamic-programming algorithm is used for addressing the long-term planning problem of a hydro-dominated system: The authors propose to compute Benders cuts in an asynchronous fashion. This is also the case in Reference [20], where the authors consider an asynchronous Benders decomposition for convex multistage stochastic programming.

Despite being successfully applied in a variety of fields, e.g., References [18,19] and the references in References [21], the classical BD is well-known to suffer from slow convergence due to the oscillatory nature of Kelley's cutting-plane method [22,23]. Regularized BDs have been proven to outperform the classical one in several problems: See Reference [24] for (convex) two-stage linear programming, Reference [25] for (nonconvex) chance-constrained problems, and Reference [26] for robust designed of stations in water distribution networks. Several types of regularization exist [25,27,28]: proximal, trust-region, and level sets. Among the regularization methods, the level bundle method [29], also known as level decomposition (LD) in two-stage programming [24], stands out for its flexibility in dealing with convex or nonconvex feasible sets, stability functions and centers, and inexact oracles [25,26,30]. Recently, asymptotically level bundle methods for convex optimization were proposed in Reference [31]. The paper presents two algorithms. The first one does not employ coordination, but it makes use of upper bounds on the Lipschitz constants of the involved functions to compute upper bounds for the problem. The second algorithm does not make use of the latter assumption but requires scarce coordination. The authors of Reference [31] focus on the convergence analysis of their proposals (suitable only for the convex setting) and present limited numerical experiments. In this work, we build on Reference [31] and extend its asynchronous algorithm with scarce coordination (Algorithm 3 of Reference [31]) to the mixed-integer setting. Moreover, we consider a more general setting in which tasks can be assigned to works in a dynamic fashion, as described in Section 3. We highlight that the convergence analysis given in Reference [31] relies strongly on elements of convex analysis such as the Smulian's theorem and the Painlevé–Kuratowski set convergence. Such key theoretical results are no longer valid in the setting of nonconvex sets, and hence the convergence analysis developed in Reference [31] does not apply to our mixed-integer setting. For this reason, the convergence analysis of our asynchronous LD must be done anew. We not only provide convergence analysis of our method but also assess its numerical performance on a test set consisting of 54 instances of two-stage UC problems with mixed-integer variables in the first stage.

We care to mention that other asynchronous bundle methods exist in the literature, but they are all designed for convex optimization problems [15,16,32]. The latter reference proposes an asynchronous proximal bundle method, whereas References [15,16] consider a trust-region variant for polyhedral functions. Our approach, which follows the lines of the extended level bundle method of Reference [30], does not require the involved functions to be polyhedral or the feasible set to be convex. As an additional advantage, our algorithm is easily implementable.

This work is organized as follows. Section 2 presents a generic formulation of our two-stage SHTUC problem. The extended asynchronous LD and its convergence analysis are presented in Sections 2.1 and 2.2, respectively. Section 3 presents more details of the considered SHTUC problem and states our case studies. Numerical experiments assessing the benefits of our proposal are given in Section 4. Finally, in Section 5, we present our final remarks.

2. Materials and Methods

We address the problem of an Independent System Operator (ISO) in a hydro-dominated system with a loose-pool market framework. The ISO decides the day-ahead commitment considering operation costs, forecast errors in wind generation, and inflows; and the usual generation and system-wide constraints. The uncertainties in wind and inflows are represented by a finite set of scenarios, $\mathcal{S}$, and the decisions are made in two stages. At the first stage, the ISO decides on the commitment of units, whereas, at the second stage, the operator determines the dispatch according to the random-variable realization. Full details on the considered stochastic hydrothermal unit-commitment (SHTUC) are given shortly. For presenting our approach, which is not limited to (stochastic) unit-commitment (UC) problems, we adopt the following generic formulation.

$$f_* := \min_{x,y} \left\{ \mathbf{c}^T x + \sum_{s \in \mathcal{S}} \mathbf{q}_s^T y_s \;\middle|\; \begin{array}{l} x \in \mathcal{X},\, \mathbf{T}x + \mathbf{W}y_s \leq \mathbf{h}_s, \\ y_s \in \mathcal{Y}_s,\, s \in \mathcal{S} \end{array} \right\}. \quad (1)$$

In this formulation, the n-dimensional vector x represents the first-stage variables with associated cost-vector, $\mathbf{c}$. The second-stage variables, y_s, and their associated costs, $\mathbf{q}_s$, depend on the scenario, $s \in \mathcal{S}$. The cost vector, $\mathbf{q}_s$, is assumed to incorporate the positive probability of scenario s. The first- and second-stage variables are coupled by constraints $\mathbf{T}x + \mathbf{W}y_s \leq \mathbf{h}_s$: $\mathbf{T}$ is the technology matrix; and $\mathbf{W}$ and $\mathbf{h}_s$ are, respectively, the recourse matrix and a vector of appropriate dimensions. While $\mathcal{X} \neq \emptyset$ is a compact possibly nonconvex, the scenario-dependent set $\mathcal{Y}_s$ is a convex polyhedron.

As previously mentioned, depending on the UC problem and number of scenarios, the mixed-integer linear programming (MILP) Problem (1) cannot be solved directly by an off-the-shelf solver. The problem is thus decomposed by making use of the recourse functions.

$$Q_s(x) := \min_{y \in \mathcal{Y}_s} \mathbf{q}_s^T y \;\text{ s.t. }\; \mathbf{W}_s y \leq \mathbf{h}_s - \mathbf{T}_s x. \quad (2)$$

It is well-known that $x \mapsto Q_s(x)$ is a non-smooth convex function of x. If the above subproblem has a solution, then a subgradient of Q_s at x can be computed by making use of a Lagrange multiplier, π_s, associated with a constraint, $\mathbf{W}_s y_s \leq \mathbf{h}_s - \mathbf{T}_s x$: $-\mathbf{T}_s^T \pi_s \in \partial Q_s(x)$. On the other hand, if the recourse function Q_s is infeasible, then the point x can be cutoff by adding a feasibility cut [5].

Let $\mathcal{P}$ be a partition of $\mathcal{S}$ into w subsets: $\mathcal{P} = \{P_1, \ldots, P_w\}$, with $P_j \neq \emptyset$ for all $j \in \{1, \ldots, w\}$, and $P_j \cap P_i = \emptyset$ for $i \neq j$. By defining $f^j(x) := \sum_{s \in P_j} Q_s(x)$, Problem (1) can be rewritten as

$$f_* = \min_{x \in \mathcal{X}} \mathbf{c}^T x + f^1(x) + \ldots + f^w(x). \quad (3)$$

In our notation, w stands for the number of workers evaluating the recourse functions. The workers $j \in \{1, \ldots, w\}$ are processes running on a single machine or multiple machines. Likewise, we define a master process—hereafter referred to only as master—to solve the master program (which is defined shortly).

2.1. The Mixed-Integer and Asynchronous Level Decomposition

For every point x_k, where k represents an iteration counter, worker j receives x_k and provides us with the first-order information on the component function f^j: the value of the function $f^j(x_k)$ and a subgradient [23] $g_k^j \in \partial f^j(x_k)$, in the two-stage setting, $g_k^j := -\sum_{s \in P_j} \mathbf{T}_s^T \pi_s$. Convexity of f^j implies that the linearization $f^j(x_k) + \langle g_k^j, x - x_k \rangle$ approximates $f^j(x)$ from below for all x. By gathering iteration indices into sets $J^j \subset \{1, 2, \ldots, k\}$ along with the iterations at which f^j were evaluated, we can construct individual cutting-plane models for functions f^j, with $j \in \{1, \ldots, w\}$: $\min_{i \in J^j} \{f^j(x_k) + \langle g_k^j, x - x_k \rangle\} \leq f^j(x)$. These models define—together with a stability center $\hat{x}_k$, a level parameter $f_k^{dev} \in \mathfrak{R}$, and a given norm $\|\cdot\|_2$—the following master program (MP)

$$\begin{cases} \min_{x,r} & \|x - \hat{x}_k\|_2 \\ \text{s.t.} & \text{possible feasibility cuts} \\ & f^j(x_i) + \langle g_i^j, x - x_i \rangle \leq r_j, \quad \forall i \in J_k^j, \forall j = 1, \ldots, w \\ & \mathbf{c}^T x + \sum_{j=1}^{w} r_j \leq f_k^{dev},\, x \in \mathcal{X}. \end{cases} \quad (4)$$

At iteration k, an MP solution is denoted by x_{k+1}. If any Q_s is infeasible at x_{k+1}, then a feasibility cut is added to the MP. We skip further details on this matter, since it is a well-known subject in the literature of two-stage programming [5]. On the other hand, if x_{k+1} (sent to a work j) is feasible for all recourse functions, Q_s, the model f^j in the MP is updated. The improvement in the model f^j is possibly based on outdated iterate $x_{a(j)}$, where a(j) < k is the iteration index of the *anterior* information provided by worker j. We care to mention that the MP can be infeasible itself depending on the level parameter f_k^{lev}. Due to the convexity of the involved functions, if the MP is infeasible, then f_k^{lev} is a valid lower bound, f_k^{low}, on f_* [30].

Without coordination, there is no reason for all workers to be called upon the same iterate. This fact precludes the computation of an upper bound, f_k^{up}, of f_*. Algorithm 2 in Reference [31] deals with this situation without resorting to coordination techniques, but it requires more assumptions on the functions f^j: upper bounds on their Lipschitz constants should be known. Since we do not make this assumption, we will need scarce coordination akin to Algorithm 3 of Reference [31] for computing upper bounds on f_*. As in Reference [31], the coordination iterates are denoted by $\bar{x}_k$. Assuming that all workers eventually respond (after an unknown time), the coordination allows them to compute the full value, $f(\bar{x}_k)$, and a subgradient, $\bar{g} \in \partial f(\bar{x}_k)$, at the coordination iterate. The function value is used to update the upper bound, f_k^{up}, as usual for level methods; the subgradient is used to update the bound L on the Lipschitz constant of f.

In our algorithm below, the coordination is implemented by two vectors of Booleans: **to-coordinate** and **coordinating**. The role of **to-coordinate**[j] is to indicate to the master that worker j will evaluate f^j on the new coordination point $\bar{x}_k$; (at that moment, **to-coordinate**[j] is set to *false*, and **coordinating**[j] is set to *true*). Similarly, **coordinating**[j] indicates to the master that worker j is responding to a coordination step, which is used to update the upper bound. When a worker j responds, it is included in the set $\mathcal{A}$ of available workers. If all workers are busy, then $\mathcal{A} = \emptyset$. Our algorithm mirrors as much as possible Algorithm 3 of Reference [31], but contains some important specificities to handle (i) mixed-integer feasible sets and (ii) extended real-valued objective functions (we do not assume that $f(x)$ is finite for all $x \in \mathcal{X}$). To handle (ii), we furnish our algorithm with a feasibility check (and addition of cuts), and for (i) we not only use a specialized solver for the MP but also change the rule for scarce coordination. The reason is that the rule of Reference [31] is only valid in the convex setting. Under nonconvexity, the coordination test $\|x_k - x_{k-1}\| < \frac{\alpha}{L}\Delta_{k-1}$ (with $\alpha \in (0,1)$ and $L \geq \|g_i\|, i = 1, \ldots, k$) implies that the following inequality (important for the convergence analysis) is jeopardized:

$$\|x_k - \hat{x}_k\|^2 \geq \|x_{k-1} - \hat{x}_k\|^2 + \left(\frac{\alpha \Delta_{k-1}}{L}\right)^2. \tag{5}$$

In the algorithm below, coordination is triggered when (5) is not satisfied and all workers have already responded on the last coordination iterate (i.e., $rr = 0$, where **rr** stands for "remaining to respond").

The assumption that the algorithm starts with a feasible point is made only for the sake of simplicity. Indeed, the initial point can be infeasible, but, in this case, Step 3 must be changed to ensure that the first computed feasible point is a coordination iterate. For the problem of interest, the feasibility check performed at line 45 amounts to verifying if $f(x_{k+1}) < \infty$. In our SHTUC, the feasibility check comprises an auxiliary problem for verifying if ramp-rate constraints would be violated by x_{k+1} and an additional auxiliary problem for checking if reservoir-volume bounds would be violated. Both problems are easily reduced to small linear-programming problems that can be solved to optimality in split seconds by off-the-shelf solvers.

Algorithm 1: Asynchronous Level Decomposition.

1. Choose a gap tolerance tol_Δ, upper bound $f_1^{up} > f_* + tol_\Delta$, lower bound $f_1^{low} < f_*$, $\alpha \in (0,1)$, $L > 0$, and x_0 a feasible point. Set $x_1 = \hat{x}_1 = x_{best} = x_0$, $\Delta_0 \leftarrow f_1^{up} - f_1^{low}$, $\hat{\Delta} \leftarrow \infty$, $rr \leftarrow 0$, $\mathcal{A} \leftarrow \{1, 2, \ldots, w\}$, $k \leftarrow 0$, $J^j \leftarrow \emptyset$ for $j \in \mathcal{A}$.
2.   for $k \leftarrow 1$ to $k+1$ do
3.   if (5) does not hold and $rr = 0$ then
4. $\bar{x}_k \leftarrow x_k$, $rr \leftarrow w$, $\bar{f} \leftarrow c^T \bar{x}_k$ and $\bar{g} \leftarrow c$
5.   for all $j \in \mathcal{A}$ do
6.   to_coordinate[j] $\leftarrow$ *false* and
7.   coordinating[j] $\leftarrow$ *true*
8.   end for
9.   for all $j \in \{1, \ldots, w\} \setminus \mathcal{A}$ do
10.   to_coordinate[j] $\leftarrow$ *true* and
11.   coordinating[j] $\leftarrow$ *false*
12.   end for
13.   end if
14.   Send x_k to all available workers $j \in \mathcal{A}$ and set $\mathcal{A} = \emptyset$
15.   Update the set $\mathcal{A}$ of idle workers and receive $(f^j(x_{a(j)}), g_{a(j)}^j)$ from workers $j \in \mathcal{A}$
16.   Update $J^j \leftarrow J^j \cup \{a(j)\}$ for all $j \in \mathcal{A}$ and set $\mathcal{R} \leftarrow \emptyset$
17.   for all $j \in \mathcal{A}$ do
18.   if coordinating[j] = *true* then
19.   coordinating[j] $\leftarrow$ *false* and $rr \leftarrow rr - 1$
20. $\bar{f} \leftarrow \bar{f} + f^j(\bar{x}_{a(j)})$ and $\bar{g} \leftarrow \bar{g} + g_{a(j)}^j$
21.   if $rr = 0$ then
22.   Set $L \leftarrow \max\{L, \|\bar{g}\|\}$
23.   if $\bar{f} < f_k^{up}$ then
24. $f_k^{up} \leftarrow \bar{f}$ and $x_{best} \leftarrow \bar{x}_k$
25.   end if
26.   end if
27.   else
28.   if to_coordinate[j] = *true* then
29.   Send $\bar{x}_k$ to worker j and set $\mathcal{R} \leftarrow \mathcal{R} \cup \{j\}$
30.   Set to_coordinate[j] $\leftarrow$ *false* and
31.   coordinating[j] $\leftarrow$ *true*
32.   end if
33.   end if
34.   end for
35.   Set $\mathcal{A} \leftarrow \mathcal{A} \setminus \mathcal{R}$
36.   Set $\Delta_k \leftarrow f_k^{up} - f_k^{low}$
37.   if $\Delta_k \leq tol_\Delta$ then stop: return x_{best} and f_k^{up} end if
38.   if $\Delta_k \leq \alpha \hat{\Delta}$ then Set $\hat{x}_k \leftarrow x_{best}$ and $\hat{\Delta} \leftarrow \Delta_k$ end if
39. $f_k^{lev} \leftarrow f_k^{up} - \alpha \Delta_k$
40.   if (4) is feasible then
41.   Get a new iterate x_{k+1} from the solution of (4)
42.   else
43.   Set $f_k^{low} \leftarrow f_k^{lev}$ and go to Step 36
44.   end if
45.   if x_{k+1} leads to infeasible subproblems then
46.   Add a feasibility cut to the MP (2) and go to Step 40
47.   end if
48.   Set $f_{k+1}^{up} \leftarrow f_k^{up}$, $f_{k+1}^{low} \leftarrow f_k^{low}$, $\hat{x}_{k+1} \leftarrow \hat{x}_k$ and $\bar{x}_{k+1} \leftarrow \bar{x}_k$
49. end for

2.2. Convergence Analysis

To analyze the convergence of the mixed-integer asynchronous computing (ASYN) level decomposition (LD) described above, we rely as much as possible on Reference [31]. However, to account for the mixed-integer nature of the feasible set, we need novel developments like the ones in Theorem 3.1 below. Throughout this section, we assume $tol_\Delta = 0$, as well as the following:

Hypothesis 1 (H1). *all the workers are responsive;*

Hypothesis 2 (H2). *algorithm generates only finitely many feasibility cuts;*

Hypothesis 3 (H3). *the workers provide bounded subgradients.*

As for H1, the assumption H2 is a mild one: H2 holds, for instance, when f is a polyhedral function, or when $\mathcal{X}$ has only finitely many points. The problem of interest satisfies both these properties, and, therefore, H2 is verified. Due to convexity of f, assumption H3 holds, e.g., if $\mathcal{X}$ is contained in an open convex set that is itself a subset of $Dom(f)$ (in this case, no feasibility cut will be generated). H3 also holds in our setting if subgradients are computed via basic optimal dual solutions of the second-stage subproblems. Under H3, we can ensure that the parameter L in the algorithm is finite.

In our analysis, we use the fact that the sequences of the optimality gap, Δ_k, and upper bound, f_k^{up}, are non-increasing by definition, and that the sequence of lower bound, f_k^{low}, is non-decreasing. More specifically, we update the lower bound only when the MP is infeasible. We count with ℓ the number of times the gap significantly decreases, meaning that the test of line 38 is triggered, and denote by $k(\ell)$ the corresponding iteration. We have the following by construction:

$$\Delta_{k(\ell+1)} \leq \alpha \Delta_{k(\ell)} \leq \alpha^2 \Delta_{k(\ell-1)} \leq \cdots \leq \alpha^\ell \Delta_1 \quad \forall \ell = 1, 2, \ldots \quad (6)$$

As in Reference [31], $k(\ell)$ denotes a critical iteration, and $x_{k(\ell)}$ denotes a critical iterate. We introduce the set of iterates between two consecutive critical iterates by $K^\ell := \{k(\ell)+1, \ldots, k(\ell+1)-1\}$. The proof of convergence of the ASYN LD consists in showing that the algorithm performs infinitely many critical iterations when $tol_\Delta = 0$. We start with the following lemma, which is a particular case of Reference [31], Lemma 3, and does not depend on the structure of $\mathcal{X}$.

Lemma 1. *Fix an arbitrary ℓ and let K^ℓ be defined as above. Then, for all $k \in K^\ell$, (a) the MP is feasible, and (b) the stability center is fixed: $\hat{x}_k = \hat{x}_{k(\ell)}$.*

Item (a) above ensures that the MP is well-defined and f_k^{low} is fixed for all $k \in K^\ell$. Note that the lower bound is updated only when the MP is found infeasible, and this fact immediately triggers the test at line 38 of the algorithm. Similarly, Algorithm 1 guarantees that the stability center remains fixed for all $k \in K^\ell$, since an updated on the stability center would imply a new critical iteration.

Theorem 1. *Assume that $\mathcal{X}$ is a compact set and that H1-H3 hold. Let $tol_\Delta = 0$ in the algorithm, and then $\lim_k \Delta_k = 0$.*

Proof of Theorem 1. By (6), we only need to show that the counter ℓ increases indefinitely (i.e., that there are infinitely many critical iterations). We obtain this by showing that, for any ℓ, the set K^ℓ is finite; for this, suppose that $\Delta_k > \Delta > 0$ for all $k \in K^\ell$. We proceed in two steps, showing the following: (i) The number of asynchronous iterations between two consecutive coordination steps is finite, and (ii) the number of coordination steps in K^ℓ is finite, as well. If case (i) were not true, then (5) and Lemma 3.1(b) would give $\|x_k - \hat{x}_{k(\ell)}\|^2 \geq \|x_{k-1} - \hat{x}_{k(\ell)}\|^2 + \left(\frac{\alpha \Delta}{L}\right)^2$, for all $k \in K^\ell$ greater than the iteration $\overline{k}$ of the last coordination iterate. Applying this inequality recursively up to $\overline{k}$, we obtain $Diam(\mathcal{X})^2 \geq \|x_k - \hat{x}_{k(\ell)}\|^2 \geq (k - \overline{k} - 1)\left(\frac{\alpha \Delta}{L}\right)^2$. However, this inequality, together with H1 and $L < \infty$

(due to H3) contradicts the fact that $\mathcal{X}$ is bounded. Therefore, item (i) holds. We now turn our attention to the item (ii): Let $s, s' \in K^\ell$ such that $s < s'$ be the iteration indices of any two coordination steps. At the moment in which $\bar{x}_{s'}$ is computed, the information $(f^j(\bar{x}_s), g_s^j)$ is available at the MP for all $j = 1, \ldots, w$. As a result of the MP definition, the following constraints are satisfied by $\bar{x}_s$:

$$f^j(\bar{x}_s) + \langle g_s^j, \bar{x}_{s'} - \bar{x}_s \rangle \leq r^j \text{ and } c^T \bar{x}_{s'} + \sum_{j=1}^{w} r^j \leq f_{s'-1}^{\text{lev}}. \tag{7}$$

By assuming these inequalities and rearranging terms, we get $f(\bar{x}_s) - f_{s'-1}^{\text{lev}} \leq \langle c + \sum_{j=1}^{w} g_s^j, \bar{x}_s - \bar{x}_{s'} \rangle \leq \Gamma \|\bar{x}_s - \bar{x}_{s'}\|$, where the constant $\infty > \Gamma \geq L \geq \|c + \sum_{j=1}^{w} g_s^j\|$ is ensured by H3. The definition of $f_{s'}^{\text{lev}} = f_{s'}^{up} - \alpha \Delta_{s'}$ and inequality $f(\bar{x}_s) \geq f_{s'}^{up}$ gives $\|\bar{x}_s - \bar{x}_{s'}\| \geq \alpha \frac{\Delta_{s'}}{\Gamma} \geq \alpha \frac{\Delta}{\Gamma} > 0$. If there was an infinite number of coordination steps inside K^ℓ, the compactness of $\mathcal{X}$ would allow us to extract a converging subsequence, and this would contradict the above inequality. The number of coordination steps inside K^ℓ is thus finite. As a conclusion of (i) and (ii), the index-set K^ℓ is hence finite, and the chain (6) concludes the proof. □

2.3. Dynamic Asynchronous Level Decomposition

In the asynchronous approach described in Algorithm 1, the component functions f^j are statically assigned to workers—worker j always evaluates the same component function j. Likewise, the usual implementation of the synchronous LD strategy is to task workers with solving fixed sets of Q_s. We call these strategies static asynchronous LD and static synchronous LD. However, as previously mentioned, such task-allocation policies might result in significant idle times—even for the asynchronous method because we need the first-order information on all f^j to compute valid bounds. To lessen the idle times, we implement dynamic-task-allocation strategies, in which component functions are dynamically assigned to workers as soon as they become available. Our dynamic allocation differs from Reference [15] because we do not use a list of iterates. To ease the understanding of the LD methods applied in this work—and to highlight their differences—we introduce a new figure: a coordinator process. The coordinator is responsible for tasking workers with functions to be evaluated. Note, however, that this additional figure is only strictly necessary in the dynamic asynchronous LD; in the other three methods, this responsibility can be taken by the master. Nonetheless, in all methods, the master has three roles: solving the MP, getting iterates, and requesting functions to be evaluated at the newly obtained iterates. By construction, in the synchronous methods, the master requests the coordinator to evaluate all functions f^j at the same iterate, and it waits until the information of the all functions has been received to continue the process. On the other hand, in the asynchronous variants, the master computes a new iterate, requests the coordinator to evaluate it on possibly not all f^j, and receives information on outdate iterates from the coordinator. Given that the master has requested an iterate x' to be evaluated in some f^j, the main difference between the static and the dynamic asynchronous methods is that, in the static form, the coordinator always sends x' to the same worker who has been previously tasked with solving f^j, while in the dynamic one, the coordinator sends x' to any available worker.

3. Modeling Details and Case Studies

The general formulation of our SHTUC is presented in (8)–(19).

$$f_* = \min \sum_{g \in \mathcal{G}} \left[\sum_{t \in \mathcal{T}} \left(CS_g \cdot a_{gt} + \sum_{s \in \mathcal{S}} C_g \cdot tg_{gts} \right) \right] + \sum_{b \in \mathcal{B}} \sum_{t \in \mathcal{T}} CL \cdot (\delta_{bt}^+ + \delta_{bt}^-) + \sum_{s \in \mathcal{S}} f_s^\omega(v) \tag{8}$$

$$\text{s.t:} \quad \sum_{o=t-TU_g+1}^{t} a_{go} \leq I_{gt}, \quad \sum_{o=t-TD_g+1}^{t} b_{go} \leq 1 - I_{gt} \tag{9}$$

$$a_{gt} - b_{gt} = I_{gt} - I_{gt-1}, z_{ht} - u_{ht} = w_{ht} - w_{ht-1} \tag{10}$$

$$z_{ht}, u_{ht}, w_{ht}, a_{gt}, b_{gt}, I_{gt} \in \{0,1\} \tag{11}$$

$$I_{gt} \cdot \underline{P}_g \leq tg_{gts} \leq I \cdot \overline{P}_g \tag{12}$$

$$tg_{gts} - tg_{gt-1s} \leq I_{gt-1} \cdot \mathbf{R}_g + (1 - I_{gt-1}) \cdot \mathbf{SU}_g \tag{13}$$

$$tg_{gt-1s} - tg_{gts} \leq I_{gt} \cdot \mathbf{R}_g + (1 - I_{gt}) \cdot \mathbf{SD}_g \tag{14}$$

$$v_{hts} - v_{ht-1s} + f^v_{hts}(q,s) + \mathbf{A}_{hts} = 0 \tag{15}$$

$$\underline{V}_h \leq v_{hts} \leq \overline{V}_h, w_{ht} \cdot \underline{Q}_h \leq q_{hts} \leq w_{ht} \cdot \overline{Q}_h, 0 \leq s_{hts} \leq \overline{S}_h \tag{16}$$

$$0 \leq hg_{hts} \leq f^{hg}_{hts}(q,s) \tag{17}$$

$$f^P_{bts}(tg, hg, \delta^+, \delta^-) + \mathbf{WG}_{bts} - \mathbf{L}_{bt} = 0 \tag{18}$$

$$\underline{\mathbf{TL}}_l \leq f^l_{lts}(tg, hg, \delta^+, \delta^-) \leq \overline{\mathbf{TL}}_l, \forall l \in \mathcal{L} \tag{19}$$

In our model, the indices and respective sets containing them are $g \in \mathcal{G}$ for thermal generators, $h \in \mathcal{H}$ for hydro plants, $b \in \mathcal{B}$ for buses, $l \in \mathcal{L}$ for transmission lines, and t and $o \in \mathcal{T}$ for periods. In (5), thermal generators' start-up costs are **CS**, and we assume that the shutdown cost is null. The thermal-generation costs are **C**; **CL** is the per-unit cost of load shedding (δ^+) and generation surplus (δ^-). Expected future-operation cost for scenario s is represented by the piecewise-affine function, $f^\omega_s(v^s): \mathbb{R}^{|\mathcal{H}|} \to \mathbb{R}$, where $v^s \in \mathbb{R}^{|\mathcal{H}|}$ are the reservoir volumes in the last period of scenario s. The first-stage decisions are thermal generators' commitment, start-up, and shutdown, respectively, I, a, and b, and their hydro counterparts (w, z, and u). Set $\mathcal{X}$ in (1) contains the feasible commitments of thermal and hydro generators in our SHTUC, and it is defined by Constraints (9)–(11). In this work, we model the statuses of hydro plants with associated binary variables only in the first 48 h, to reduce the computational burden. For the remaining periods, the hydro plants are modeled only with continuous variables. The minimum up-time Constraint (9) ensures that, once turned on, thermal generator g remains on for at least $\mathbf{TU}_g$ periods. Likewise, the minimum downtime in (9) requires that once g has been turned off, it must remain off for at least $\mathbf{TD}_g$ periods. Constraints (10) guarantee the satisfaction of logical relations of status, start-up, and shutdown for thermal and hydro plants. The sets $\mathcal{Y}$s are defined by (12)–(19). Constraints (12) are the usual limits on thermal generation tg; (13) and (14) are the up and down ramp-rate limits, and the start-up and shutdown requirements of generators g. Equation (15) is the mass balance of the hydro plant h's reservoir. The $\mathbf{A}_{hts}$ is the inflow to reservoir h in period t of scenario s. Moreover, the affine function $f^v_{hts}(q,s): \mathbb{R}^{2 \cdot |\mathcal{H}| \cdot |\mathcal{T}| \cdot |\mathcal{S}|} \to \mathbb{R}$ maps the inflow to h's reservoir in period t of scenario s given the vectors of turbine discharge q and spillage s. The constraints in (16) are the limits on reservoir volume, v, turbine discharge, q, and spillage, s. In (17), the piecewise-affine function $f^{hg}_{hts}(q,s): \mathbb{R}^{2 \cdot |\mathcal{H}| \cdot |\mathcal{T}| \cdot |\mathcal{S}|} \to \mathbb{R}$ bounds the hydropower generation hg_{hts} of plant h. We use the classical DC network model: Equation (18) is the bus power balance, where the linear function $f^P_{bts}(tg, hg, \delta^+, \delta^-): \mathbb{R}^{|\mathcal{T}| \cdot |\mathcal{S}| \cdot (|\mathcal{G}|+|\mathcal{H}|+2 \cdot |\mathcal{B}|)} \to \mathbb{R}$ maps the controlled generation at each bus into the power injection at bus b, $\mathbf{WG}_{bts}$ is the wind generation at bus b, and $\mathbf{L}_{bt}$ is the corresponding load at b. Lastly, (19) are the limits on the flow of transmission line l in period t and scenario s, defined by the affine function $f^l_{lst}(tg, hg, \delta^+, \delta^-): \mathbb{R}^{|\mathcal{T}| \cdot |\mathcal{S}| \cdot (|\mathcal{G}|+|\mathcal{H}|+2 \cdot |\mathcal{B}|)} \to \mathbb{R}$.

We assess our algorithm on a 46-bus system with 11 thermal plants, 16 hydro plants, 3 wind farms, and 95 transmission lines. The system's installed capacity is 18,600 MW, from which 18.9% is due to thermal plants, hydro plants represent 68.1%, and wind farms have a share of 13%. We consider a one-week-long planning horizon with hourly discretization. Thus, a one-scenario instance of our

SHTUC would have 7848 binary variables and 5315 constraints at the first stage; and 36,457 continuous variables and 100,949 constraints for each scenario in the second stage. Furthermore, the weekly peak load in the baseline case is 11,204 MW—nearly 60.2% of the installed capacity. The hydro plants are distributed over two basins and include both run-of-river ones and plants with reservoirs capable of regularization. Further information about the system can be found in the multimedia files attached.

The uncertainty comes from wind generation and the inflows. In all tests, we use a scenario set with 256 scenarios. To assess how our algorithm performs in distinct scenario sets, three sets (A, B, and C) are considered. Moreover, we use three initial useful-reservoir-volume levels: 40%, 50%, and 70%. The impact of different load levels on the performance of our algorithms is analyzed through three load levels: low (L), moderate (M), and high (H). Level H is our baseline case regarding load. Levels M and L have the same load profile as H's, but with all loads multiplied by factors of 0.9 and 0.8, respectively. Lastly, to investigate how our algorithm's convergence rate is affected by different choices of initial stability centers, we implement two strategies for obtaining the initial stability center. In both strategies, we solve an expected-value problem, as defined in Reference [5]. In the first one, we use the classical Benders decomposition (BD) with a coarse relative-optimality-gap tolerance of 10% to get a, possibly, low-quality stability center (LQSC). To obtain the stability center of hopefully high quality, which we refer to as high-quality stability center (HQSC), we solve the expected-value problem directly with Gurobi 8.1.1 [33] with a relative-optimality-gap tolerance of 1%. The time limit for obtaining the initial stability centers LQSC and HQSC is set to 5 min. Additionally, the computing setting consists of seven machines of two types: 4 of them have 128 GB of RAM and two Xeon E5-2660 v3 processors with 10 cores clocking at 2.6 GHz; the other 3 machines have 32 GB of RAM and two Xeon X5690 processors with cores cores clocking at 3.47 GHz. All machines are in a LAN with 1-Gbps network interfaces. We test two machine combinations. In the first one, in Combination 1, there are four 20-core machines and one with 12 cores. In Combination 2, we replace one machine with 20 cores by 2 with 12 cores. Regardless of the combination, one 12-core machine is defined as the head node, where only the master is launched. Except for the master—for which Gurobi can take up to 10 cores—for all other processes, i.e., the workers, Gurobi is limited to computing on a single core.

Our computing setting is composed of machines with different configurations. Naturally, solving the same component function in two distinct machines may result in different outputs—and different runtimes. Consequently, the path taken by the MP across iterations might change significantly between experiments on the same data. More specifically to asynchronous methods, the varying order of information arrival to the MP may also yield different convergence rates. Hence, to reduce the effect of these seemingly random behaviors, we conducted 5 experiments for each problem instance. Therefore, our testbed $\mathcal{E}$ is defined as $\mathcal{E} = \{40, 50, 70\} \times \{A, B, C\} \times \{L, M, H\} \times \{LQ\text{-}SC, HQ\text{-}SC\} \times \{\text{Trial } 1, \ldots, \text{Trial } 5\} \times \{\text{Combination 1, Combination 2}\}$—we have 54 problems and 540 experiments. In all instances in $\mathcal{E}$, we divide $\mathcal{S}$ into 16 subsets. Thus, following our previous definitions, $w = 16$ and any subset $\mathcal{P}_j$ is such that $|\mathcal{P}_j| = 16$. Additionally, we set a relative-optimality-gap tolerance of 1% and a time limit of 30 min for all instances in $\mathcal{E}$. Gurobi 8.1.1 is used to solve the MILP MP and the component functions (linear-programming problems) that form the subproblem. The inter-process communication is implemented with mpi4py and Microsoft MPI v10.0.

4. Results

In this section, the methods are analyzed based on their computing-time performances. We focus on this metric because our results have not shown significant differences among the methods for other metrics, e.g., optimality gap and upper bounds. In addition to analyzing averages of the metric, we use the well-known performance profile [34]. Multimedia files containing the main results for the set $\mathcal{E}$ are attached to this work.

Figure 1 presents the performance profiles of the methods considering the experiments $\mathcal{E}$. In Figure 1, $\rho(\tau)$ and τ are, respectively, the probability that the performance ratio of a given method is within a factor τ of the best ratio, as in Reference [34]. Applying the classical Benders decomposition

(BD) on the set {40, 50, 70} × {A} × {L, M, H} × {Combination 1} results in the convergence only of the problem in {70} × {A} × {M} × {Combination 1}, for which BD converges to a 1%-optimal solution in 1281.42 s. Thus, it is reasonable to expect that the classical BD would also perform poorly for the remaining experiments $\mathcal{E}$.

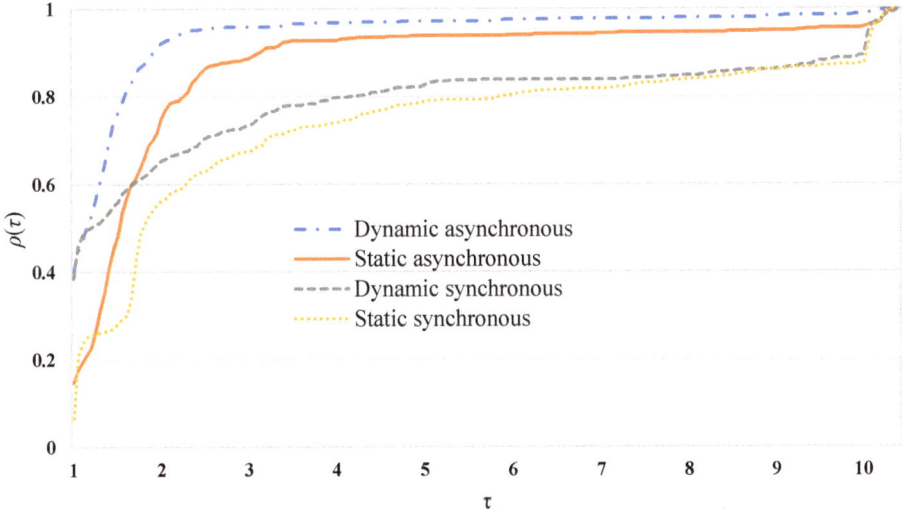

Figure 1. Performance profiles over the set $\mathcal{E}$.

In Figure 1, we see that the dynamic asynchronous LD outperforms all other methods for most instances $\mathcal{E}$. Its performance ratio is within a factor of 2 from the best ratio for about 500 instances (about 92% of the total). Moreover, the static asynchronous LD has a reasonable overall performance—it is within a factor of 2 from the best ratio for more than 400 instances. Moreover, we see that the dynamic-allocation strategy provides significant improvements for both the asynchronous and synchronous LD approaches. The dynamic synchronous LD converges faster than its static counterpart for most of the experiments. Figures 2 and 3 show the performance profiles considering only instances in $\mathcal{E}$ with machine Combinations 1 and 2, respectively.

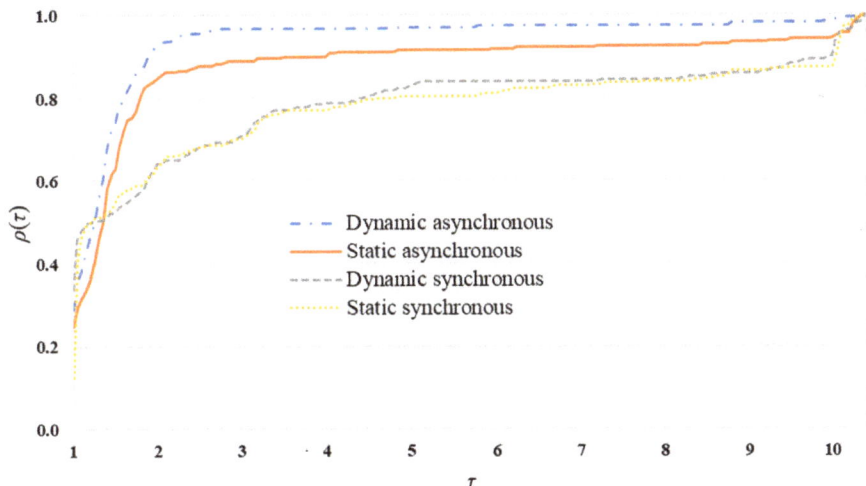

Figure 2. Performance profiles for the instances with machine Combination 1.

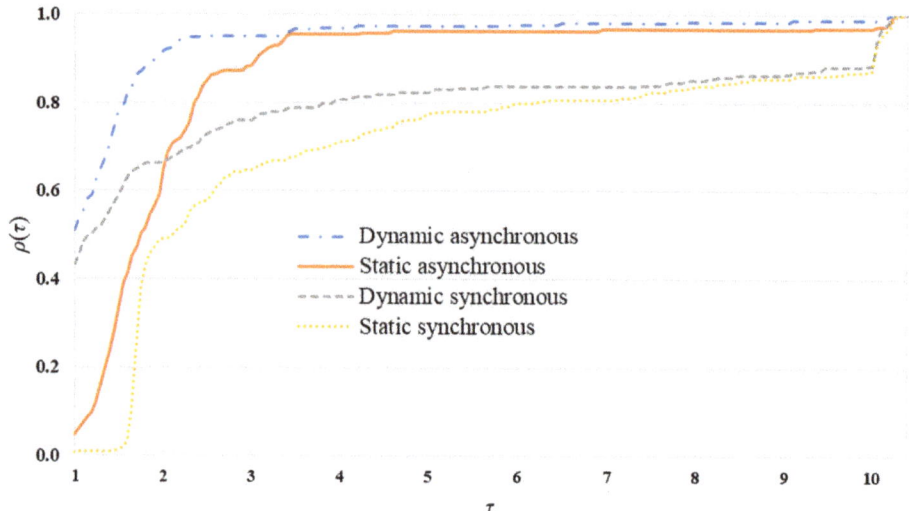

Figure 3. Performance profiles for the instances with machine Combination 2.

Figure 2 illustrates that, for a distributed setting in which workers are deployed on machines with identical characteristics, the performances of the methods with dynamic allocation and those with static allocation are similar. Nonetheless, we see that the asynchronous methods still outperform the synchronous LD for most experiments.

In contrast to Figure 2, Figure 3 shows that the dynamic-allocation strategy provides significant time savings for the instances in $\mathcal{E}$ with Machine Combination 2. This is due to the great imbalance between the different machines in Combination 2—machines with processors Xeon E5-2660 v3 are much faster than those with processors Xeon X5690.

Table 1 gives the average wall-clock computing times over subsets of $\mathcal{E}$. From this table, we see that the relative average speed-up of the dynamic and static asynchronous LD over the entire set $\mathcal{E}$ w.r.t. The static synchronous LD are 54% and 29%, respectively—considering the dynamic synchronous LD, the speed-ups are 45% and 16%, respectively. Moreover, we see that the time savings are more significant for harder-to-solve instances, e.g., instances with high load and/or low-quality initial stability centers. Additionally, Table 1 shows that the dynamic asynchronous LD provides considerable reductions in the standard deviations of the elapsed computing times, in comparison with the other methods. For example, for the problems with high load level (H), the dynamic asynchronous LD has a standard deviation of about 16%, 13%, and 27% smaller than that of the static asynchronous LD, dynamic synchronous LD, and static synchronous LD, respectively.

Based on the data from Table 1, we can compute the speed-up provided by our proposed dynamic ASYN LD w.r.t., and the other three variants are considered here. To better appreciate such speed-ups, we show them in Table 2, where we see that the proposed ASYN LD provides consistent speed-ups over the entire range of operating conditions considered here.

The advantages of the asynchronous methods are made clearer in Figure 4, where we see that not only the asynchronous methods provide (on average) better running times but also present significantly less variation among the problems in $\mathcal{E}$. The latter is relevant in the day-to-day operations of ISOs, since, if there are stochastic hydrothermal unit-commitment (SHTUC) cases that take significantly more time to be solved than the expected, subsequent operation steps that depend on the results of the SHTUC might be affected. Take, for instance, the case from the Midcontinent Independent System Operator reported in Reference [3], where the (deterministic) UC is reported to have solution times varying from just 50 to over 3600 s. Such variation can be problematic in the day-to-day operation of power systems since it may disrupt tightly scheduled operations. Naturally, methods that can reduce such variance and still produce high-quality solutions in reasonable times are appealing.

Table 1. Average elapsed time and standard deviation in seconds.

	Asynchronous		Synchronous	
	Dynamic	Static	Dynamic	Static
40	135 (340)	240 (497)	364 (569)	338 (490)
50	130 (279)	222 (414)	195 (293)	267 (404)
70	127 (249)	139 (242)	161 (157)	250 (300)
A	143 (280)	226 (418)	187 (270)	216 (303)
B	137 (340)	192 (406)	305 (515)	347 (530)
C	112 (247)	184 (377)	229 (335)	291 (340)
L	102 (195)	157 (317)	172 (398)	201 (396)
M	108 (179)	112 (111)	142 (126)	205 (210)
H	182 (425)	333 (585)	405 (492)	449 (506)
HQSC	105 (202)	129 (193)	156 (258)	159 (118)
LQSC	156 (357)	272 (523)	324 (473)	411 (534)
Combination 1	120 (282)	207 (447)	227 (387)	243 (416)
Combination 2	141 (301)	194 (348)	253 (393)	327 (393)

The rows indicate that the average elapsed times and standard deviation given in parentheses are computed considering only the instances in $\mathcal{E}$ with the parameter given in the column 1. For example, the averages and respective standard deviations in row 3 are computed considering all experiments for which the initial useful-reservoir-volume level is 40%. Likewise, rows 4 and 7 provide the averages over instances with scenario set A and load level L, respectively. In rows 10 and 11, HQSC and LQSC stand for high-quality stability center and low-quality stability center, respectively.

Table 2. Speed-ups in % provided by the asynchronous computing (ASYN) with respect to the level decomposition (LD)

	Static SYN LD	Dynamic SYN LD	Static ASYN LD
40	60	63	44
50	51	33	41
70	49	21	9
A	34	23	37
B	61	55	29
C	61	51	39
L	49	41	35
M	47	24	4
H	59	55	45
HQSC	34	33	19
LQSC	62	52	42
Combination 1	50	47	42
Combination 2	57	44	27

As in Table 1, the rows indicate that the average speed-up computed considering only the instances in $\mathcal{E}$ with the parameter given in the column 1. Moreover, the columns indicate the method the speed-up is computed for. For example, column Static SYN (synchronous computing) LD gives the speed-ups provided by the ASYN LD over instances in the first column w.r.t. to the static synchronous level decomposition.

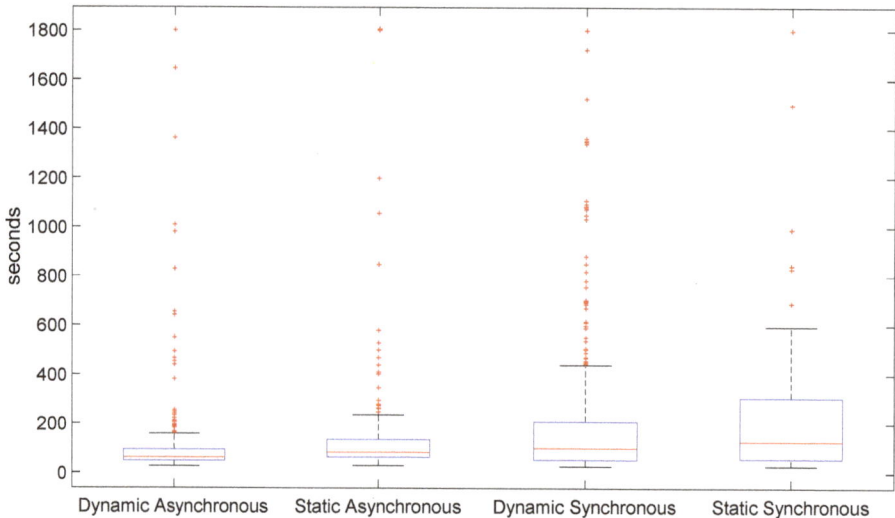

Figure 4. Boxplot of the methods over the set $\mathcal{E}$.

5. Conclusions

In this work, we present an extension of the asynchronous level decomposition of Reference [31] in a Benders-decomposition framework. We show a convergence analysis of our algorithm, proving that it converges to an optimal solution, if one exists, in finite-many iterations. Our experiments are conducted on an extensive testbed from a real-life-size system. The results show that the proposed asynchronous algorithm outperforms its synchronous counterpart in most of the problems and provides significant time savings. Moreover, we show that the improvements provided by the asynchronous methods over the synchronous ones are even more evident in a distributed-computing setting with machines of different computational powers. Additionally, we show that the asynchronous method is further enhanced by implementing a dynamic-task-allocation strategy.

Author Contributions: Conceptualization, B.C., E.C.F., and W.d.O.; methodology, B.C., E.C.F., and W.d.O.; software, B.C.; validation, B.C.; formal analysis, E.C.F. and W.d.O.; investigation, B.C., E.C.F., and W.d.O.; resources, B.C. and E.C.F.; data curation, B.C.; writing—original draft preparation, B.C., E.C.F., and W.d.O.; writing—review and editing, B.C., E.C.F., and W.d.O.; visualization, B.C., E.C.F., and W.d.O.; supervision, E.C.F. and W.d.O. All authors have read and agreed to the published version of the manuscript.

Funding: The third author acknowledges financial support from the Gaspard-Monge program for Optimization and Operations Research (PGMO) project "Models for planning energy investment under uncertainty".

Conflicts of Interest: The authors declare no conflict of interest.

References

1. Zheng, Q.P.; Wang, J.; Liu, A.L. Stochastic Optimization for Unit Commitment—A Review. *IEEE Trans. Power Syst.* **2015**, *30*, 1913–1924. [CrossRef]
2. Santos, T.N.; Diniz, A.L.; Saboia, C.H.; Cabral, R.N.; Cerqueira, L.F. Hourly Pricing and Day-Ahead Dispatch Setting in Brazil: The Dessem Model. *Electr. Power Syst. Res.* **2020**, *189*, 106709. [CrossRef]
3. Chen, Y.; Pan, F.; Holzer, J.; Rothberg, E.; Ma, Y.; Veeramany, A. A High Performance Computing Based Market Economics Driven Neighborhood Search and Polishing Algorithm for Security Constrained Unit Commitment. *IEEE Trans. Power Syst.* **2020**, 1. [CrossRef]
4. Tahanan, M.; Van Ackooij, W.; Frangioni, A.; Lacalandra, F. Large-Scale Unit Commitment under Uncertainty. *4OR* **2015**, *13*, 115–171. [CrossRef]
5. Birge, J.R.; Louveaux, F. *Introduction to Stochastic Programming*; Springer Series in Operations Research and Financial Engineering; Springer New York: New York, NY, USA, 2011. [CrossRef]

6. Håberg, M. Fundamentals and Recent Developments in Stochastic Unit Commitment. *Int. J. Electr. Power Energy Syst.* **2019**, *109*, 38–48. [CrossRef]
7. Sagastizábal, C. Divide to Conquer: Decomposition Methods for Energy Optimization. *Math. Program.* **2012**, *134*, 187–222. [CrossRef]
8. Benders, J.F. Partitioning Procedures for Solving Mixed-Variables Programming Problems. *Numer. Math.* **1962**, *4*, 238–252. [CrossRef]
9. Bagirov, A.M.; Ozturk, G.; Kasimbeyli, R. A Sharp Augmented Lagrangian-Based Method in Constrained Non-Convex Optimization. *Optim. Methods Softw.* **2019**, *34*, 462–488. [CrossRef]
10. Bertsekas, D.P.; Tsitsiklis, J.N. *Parallel and Distributed Computation: Numerical Methods*; Athena Scientific: Nashua, NH, USA, 2015.
11. Khanabadi, M.; Fu, Y.; Gong, L. A Fully Parallel Stochastic Multiarea Power System Operation Considering Large-Scale Wind Power Integration. *IEEE Trans. Sustain. Energy* **2018**, *9*, 138–147. [CrossRef]
12. Papavasiliou, A.; Oren, S.S.; Rountree, B. Applying High Performance Computing to Transmission-Constrained Stochastic Unit Commitment for Renewable Energy Integration. *IEEE Trans. Power Syst.* **2015**, *30*, 1109–1120. [CrossRef]
13. Kargarian, A.; Fu, Y.; Li, Z. Distributed Security-Constrained Unit Commitment for Large-Scale Power Systems. *IEEE Trans. Power Syst.* **2015**, *30*, 1925–1936. [CrossRef]
14. Kargarian, A.; Mehrtash, M.; Falahati, B. Decentralized Implementation of Unit Commitment With Analytical Target Cascading: A Parallel Approach. *IEEE Trans. Power Syst.* **2018**, *33*, 3981–3993. [CrossRef]
15. Kim, K.; Petra, C.G.; Zavala, V.M. An Asynchronous Bundle-Trust-Region Method for Dual Decomposition of Stochastic Mixed-Integer Programming. *SIAM J. Optim.* **2019**, *29*, 318–342. [CrossRef]
16. Kim, K.; Anitescu, M.; Zavala, V.M. An Asynchronous Decomposition Algorithm for Security Constrained Unit Commitment Under Contingency Events. In *Proceedings of the 2018 Power Systems Computation Conference (PSCC)*; IEEE: Piscataway, NJ, USA, 2018; pp. 1–8. [CrossRef]
17. Aravena, I.; Papavasiliou, A. A Distributed Asynchronous Algorithm for the Two-Stage Stochastic Unit Commitment Problem. In *Proceedings of the 2015 IEEE Power & Energy Society General Meeting*; IEEE: Piscataway, NJ, USA, 2015; pp. 1–5. [CrossRef]
18. Santos, T.N.; Diniz, A.L.; Borges, C.L.T. A New Nested Benders Decomposition Strategy for Parallel Processing Applied to the Hydrothermal Scheduling Problem. *IEEE Trans. Smart Grid* **2017**, *8*, 1504–1512. [CrossRef]
19. Pinto, R.J.; Borges, C.T.; Maceira, M.E.P. An Efficient Parallel Algorithm for Large Scale Hydrothermal System Operation Planning. *IEEE Trans. Power Syst.* **2013**, *28*, 4888–4896. [CrossRef]
20. Moritsch, H.W.; Pflug, G.C.; Siomak, M. Asynchronous Nested Optimization Algorithms and Their Parallel Implementation. *Wuhan Univ. J. Nat. Sci.* **2001**. [CrossRef]
21. Rahmaniani, R.; Crainic, T.G.; Gendreau, M.; Rei, W. The Benders Decomposition Algorithm: A Literature Review. *Eur. J. Oper. Res.* **2017**, *259*, 801–817. [CrossRef]
22. Kelley, J.E., Jr. The Cutting-Plane Method for Solving Convex Programs. *J. Soc. Ind. Appl. Math.* **1960**. [CrossRef]
23. Bagirov, A.; Karmitsa, N.; Mäkelä, M.M. *Introduction to Nonsmooth Optimization*; Springer International Publishing: Cham, Switzerland, 2014. [CrossRef]
24. Wolf, C.; Fábián, C.I.; Koberstein, A.; Suhl, L. Applying Oracles of On-Demand Accuracy in Two-Stage Stochastic Programming—A Computational Study. *Eur. J. Oper. Res.* **2014**, *239*, 437–448. [CrossRef]
25. Van Ackooij, W.; Frangioni, A.; de Oliveira, W. Inexact Stabilized Benders' Decomposition Approaches with Application to Chance-Constrained Problems with Finite Support. *Comput. Optim. Appl.* **2016**. [CrossRef]
26. Bonvin, G.; Demassey, S.; de Oliveira, W. Robust Design of Pumping Stations in Water Distribution Networks. In *Advances in Intelligent Systems and Computing*; Springer: Berlin/Heidelberg, Germany, 2020; Volume 991, pp. 957–967. [CrossRef]
27. Mäkelä, M.M.; Karmitsa, N.; Bagirov, A. *Subgradient and Bundle Methods for Nonsmooth Optimization*; Springer: Dordrecht, The Netherlands, 2013; pp. 275–304. [CrossRef]
28. Bagirov, A.M.; Gaudioso, M.; Karmitsa, N.; Mäkelä, M.M.; Taheri, S. (Eds.) *Numerical Nonsmooth Optimization*; Springer International Publishing: Cham, Switzerland, 2020. [CrossRef]
29. Lemaréchal, C.; Nemirovskii, A.; Nesterov, Y. New Variants of Bundle Methods. *Math. Program.* **1995**. [CrossRef]
30. De Oliveira, W. Regularized Optimization Methods for Convex MINLP Problems. *TOP* **2016**, *24*, 665–692. [CrossRef]
31. Iutzeler, F.; Malick, J.; de Oliveira, W. Asynchronous Level Bundle Methods. *Math. Program.* **2019**. [CrossRef]

32. Fischer, F.; Helmberg, C. A Parallel Bundle Framework for Asynchronous Subspace Optimization of Nonsmooth Convex Functions. *SIAM J. Optim.* **2014**, *24*, 795–822. [CrossRef]
33. Gurobi Optimization LLC. *Gurobi Optimizer Reference Manual*; Gurobi Optimization LLC: Beaverton, OR, USA, 2018.
34. Dolan, E.D.; Moré, J.J. Benchmarking Optimization Software with Performance Profiles. *Math. Program.* **2002**, *91*, 201–213. [CrossRef]

© 2020 by the authors. Licensee MDPI, Basel, Switzerland. This article is an open access article distributed under the terms and conditions of the Creative Commons Attribution (CC BY) license (http://creativecommons.org/licenses/by/4.0/).

Article

On a Nonsmooth Gauss–Newton Algorithms for Solving Nonlinear Complementarity Problems

Marek J. Śmietański

Faculty of Mathematics and Computer Science, University of Lodz, Banacha 22, 90-238 Łódź, Poland; marek.smietanski@wmii.uni.lodz.pl

Received: 25 June 2020; Accepted: 31 July 2020; Published: 4 August 2020

Abstract: In this paper, we propose a new version of the generalized damped Gauss–Newton method for solving nonlinear complementarity problems based on the transformation to the nonsmooth equation, which is equivalent to some unconstrained optimization problem. The B-differential plays the role of the derivative. We present two types of algorithms (usual and inexact), which have superlinear and global convergence for semismooth cases. These results can be applied to efficiently find all solutions of the nonlinear complementarity problems under some mild assumptions. The results of the numerical tests are attached as a complement of the theoretical considerations.

Keywords: Gauss–Newton method; nonsmooth equations; nonsmooth optimization; nonlinear complementarity problem; B-differential; superlinear convergence; global convergence

1. Introduction

Let $F : R^n \to R^n$ and let F_i, $i = 1, ..., n$ denote the components of F. The nonlinear complementarity problem (NCP) is to find $x \in R^n$ such that

$$x \geq 0, F(x) \geq 0 \text{ and } x^T F(x) = 0. \tag{1}$$

The ith component of a vector x is represented by x_i. Solving (1) is equivalent to solving a nonlinear equation $G(x) = 0$, where the operator $G : R^n \to R^n$ is defined by

$$G(x) = \begin{bmatrix} \varphi(x_1, F_1(x)) \\ ... \\ \varphi(x_n, F_n(x)) \end{bmatrix}$$

with some special function φ. Function φ may have one of the following forms:

$$\begin{aligned}
\varphi_1(a,b) &= \min\{a,b\}; \\
\varphi_2(a,b) &= \sqrt{a^2+b^2} - a - b; \\
\varphi_3(a,b) &= \theta(|a-b|) - \theta(a) - \theta(b),
\end{aligned}$$

where $\theta : R \to R$ is any strictly increasing function with $\theta(0) = 0$, see [1].

The (NCP) problem is one of the fundamental problems of mathematical programming, operations research, economic equilibrium models, and in engineering sciences. A lot of interesting and important applications can be found in the papers of Harker and Pang [2] and Ferris and Pang [3]. We can find the most essential applications in:

- engineering—optimal control problems, contact or structural mechanics problems, structural design problems, or traffic equilibrium problems,
- equilibrium modeling—general equilibrium (in production or consumption), invariant capital stock, or game-theoretic models.

We borrow a technique used in solving some smooth problems. If g is a merit function of G, i.e., $g(x) = \frac{1}{2} G(x)^T G(x)$, then any stationary point of $g(x)$ is a least-squares solution of the equation $G(x) = 0$. Then, algorithms for minimization are equivalent to algorithms for solving equations. The usual Gauss–Newton method (known also as the differential corrections method), presented by Ortega and Rheinboldt [4] in the smooth case, has the form

$$x^{(k+1)} = x^{(k)} - \left[G'(x^{(k)})^T G'(x^{(k)}) \right]^{-1} G'(x^{(k)})^T G(x^{(k)}). \qquad (2)$$

Local convergence properties of the Gauss–Newton method was discussed by Chen and Li [5], but only for some smooth case. The Levenberg–Marquardt method is also considered, which is a modified Gauss–Newton method, in some papers, e.g., [6] or [7]. Moreover, some comparison of semismooth algorithms for solving (NCP) problems has been made in [8].

In practice, we may also consider the damped Gauss–Newton method

$$x^{(k+1)} = x^{(k)} - \omega_k \left[G'(x^{(k)})^T G'(x^{(k)}) + \lambda_k I \right]^{-1} G'(x^{(k)})^T G(x^{(k)}) \qquad (3)$$

with parameters ω_k and λ_k. Parameter ω_k may be chosen to ensure suitable decrease of g. If λ_k is positive for all k, then the inverse matrix in (3) always exists because $G'(x^{(k)})^T G'(x^{(k)})$ is a symmetric and positive semidefinite matrix. The method (3) has the important advantage: the search direction always exists, even if $G'(x)$ is singular. Naturally, in the case of nonsmooth equations, some additional assumptions are needed to allow the use of some line search strategies and to ensure the global convergence. Because, in some cases, a function G is nondifferentiable, so the equation $G(x) = 0$ will be nonsmooth, whereby the method (3) may be useless. Some version of the Gauss–Newton method for solving complementarity problems was also introduced by Xiu and Zhang [9] for generalized problems, but only for linear ones. Thus, for solving nonsmooth and nonlinear problems, we propose two new versions of a damped Gauss–Newton algorithm based on B-differential. The usual generalized method is a relevant extension of the work by Subramanian and Xiu [10] for a nonsmooth case. In turn, an inexact version is related to the traditional approach, which was widely studied, e.g., in [11]. In recent years, various versions of the Gauss–Newton method were discussed, although most frequently for solving nonlinear least-squares problems, e.g., in [12,13].

The paper is organized as follows: in the next section, we review some notions needed, such as B-differential, BD-regularity, semismoothness, etc. (Section 2.1). Next, we propose a new optimization problem-based methods for the NCP, transforming the NCP into an unconstrained minimization problem by employing a function φ_3 (Section 2.2). We state its global convergence and superlinear convergence rate under appropriate conditions. In Section 3, we present the results of numerical tests.

2. Materials and Methods

2.1. Preliminaries

If F is Lipschitz continuous, the Rademacher's theorem [14] implies that F is almost everywhere differentiable. Let the set of points, where F is differentiable, be denoted by D_F. Then, the B-differential (the Bouligand differential) of F at x (introduced in [15]) is

$$\partial_B F(x) = \left\{ \lim_{x^{(n)} \to x} F'\left(x^{(n)}\right), x^{(n)} \in D_F \right\},$$

where $F'(x)$ denotes the usual Jacobian of F at x. The generalized Jacobian of F at x in the sense of Clarke [14] is

$$\partial F(x) = \text{conv} \partial_B F(x)$$

We say that F is BD-regular at x, if F is locally Lipschitz at x and if all $V \in \partial_B F(x)$ are nonsingular (regularity on account of B-differential). Qi proved (Lemma 2.6, [15]) that, if F is BD-regular at x, then a neighborhood N of x and a constant $C > 0$ exist such that, for any $y \in N$ and $V \in \partial_B F(y)$, V is nonsingular and

$$\left\| V^{-1} \right\| \leq C.$$

Throughout this paper, $\|\cdot\|$ denotes the 2-norm.

The notion of semismoothness was originally introduced for functionals by Mifflin [16]. The following definition is taken from Qi and Sun [17]. A function F is semismooth at a point x, if F is locally Lipschitzian at x and

$$\lim_{V \in \partial F(x+th'), h' \to h, t \downarrow 0} Vh'$$

exists for any $h \in R^n$. F is also said semismooth at x, if it is directionally differentiable at x and

$$Vh - F'(x, h) = o(\|h\|).$$

Scalar products and sums of semismooth functions are still semismooth functions. Piecewise smooth functions and maximum of a finite number of smooth functions are also semismooth. The semismoothness is the almost usually seen assumption on F in papers dealing with nonsmooth equations because it implies some important properties for convergence analysis of methods in nonsmooth optimization.

If for any $V \in \partial F(x+h)$, as $h \to 0$

$$Vh - F'(x, h) = O\left(\|h\|^{1+p}\right),$$

where $0 < p \leq 1$, then we say F is p-order semismooth at x. Clearly, p-order semismoothness implies semismoothness. If $p = 1$, then the function F is called strongly semismooth. Piecewise C^2 functions are examples of strongly semismooth functions.

Qi and Sun [17] remarked that, if F is semismooth at x, then, for any $h \to 0$

$$F(x+h) - F(x) - F'(x; h) = o(\|h\|),$$

and, if F is p-order semismooth at x, then for any $h \to 0$

$$F(x+h) - F(x) - F'(x; h) = O\left(\|h\|^{1+p}\right).$$

Remark 1. *Strong semismoothness of the appropriate function usually implies quadratic convergence of method instead of the superlinear one for semismooth function.*

In turn, Pang and Qi [18] proved that semismoothness of F at x implies that

$$\sup_{V \in \partial F(x+h)} \{F(x+h) - F(x) - Vh\} = o(\|h\|).$$

Moreover, if F is p-order semismooth at x, then

$$\sup_{V\in\partial F(x+h)} \{F(x+h) - F(x) - Vh\} = O\left(\|h\|^{1+p}\right).$$

2.2. The Algorithm and Its Convergence

Consider nonlinear equation $G(x) = 0$ defined by φ_3. The equivalence of solving this equation and problem (NCP) is described by the following theorem:

Theorem 1 (Mangasarian [1]). *Let θ be any strictly increasing function from R into R, that is, $a > b \Leftrightarrow \theta(a) > \theta(b)$, and let $\theta(0) = 0$. Then, x solves the complementarity problem (1) if and only if*

$$\theta(|F_i(x) - x_i|) - \theta(F_i(x)) - \theta(x_i) = 0, \ i = 1, 2, \ldots, n. \tag{4}$$

For the convenience, denote

$$G_i(x) := \theta(|F_i(x) - x_i|) - \theta(F_i(x)) - \theta(x_i) \tag{5}$$

for $i = 1, 2, \ldots, n$.

We assume that the function θ in Theorem 1 has the form

$$\theta(\xi) = \xi |\xi|.$$

Let $G(x)$ be the associated function. We define function g in the following way:

$$g(x) = \frac{1}{2}\|G(x)\|^2,$$

which allows for solving system $G(x) = 0$ based on solving the nonlinear least-square problem

$$\min_x g(x). \tag{6}$$

Let us note that x^* solves $G(x) = 0$ if and only if it is a stationary point of g. Thus, from Theorem 1, x^* solves (1).

Remark 2. *On the other hand, the first-order optimality conditions for problem (6) are equivalent to the nonlinear system*

$$\nabla g(x) = G'(x)^T G(x) = 0,$$

where ∇g is the gradient of g, provided G is differentiable and G' is the Jacobian matrix of G.

The continuous differentiability of the merit function g for some kind of nonsmooth functions was established by Ulbrich in the following lemma:

Lemma 1 (Ullbrich, [19]). *Assume that the function $G : R^n \supset D \to R^n$ is semismooth, or, stronger, p-order semismooth, $0 < p \leq 1$, then the merit function $\frac{1}{2}\|G(x)\|^2$ is continuously differentiable on D with gradient $\nabla g(x) = V^T G(x)$, where $V \in \partial G(x)$ is arbitrary.*

Lemma 2. *For any $x \in R^n$, let $A_x = V_x^T V_x$, where $V_x \in \partial_B G(x)$. Suppose that $\nabla g(x) \neq 0$. Then, given $\lambda > 0$, the direction d given by*

$$(A_x + \lambda I)d = \nabla g(x)$$

is an ascent direction for g. In particular, there is a positive w such that $g(x - wd) < g(x)$.

Proof. There exist constants $\beta \geq 0$ and $\gamma > 0$ such that

$$\beta \|h\|^2 \leq h^T A_x h \leq \gamma \|h\|^2 \text{ for all } h \in R^n,$$

because A_x defined as $V_x^T V_x$ is symmetric and positive semidefinite.

It follows that

$$(\beta + \lambda) \|h\|^2 \leq h^T (A_x + \lambda I) h \leq (\gamma + \lambda) \|h\|^2 \text{ for all } h \in R^n.$$

Since $\nabla g(x) \neq 0, d \neq 0$. If we take $h = d$, we obtain

$$d^T \nabla g(x) \geq (\beta + \lambda) \|d\|^2 > 0.$$

It follows that $\nabla g(x) d > 0$ and that d is a ascent direction for g (Section 8.2.1 in [4]). □

Now, we present the generalized version of the damped Gauss–Newton method for solving the nonlinear complementarity problem.

Algorithm 1: The damped Gauss-Newton method for solving NCP

Let $\beta, \delta \in (0, 1)$ be given. Let $x^{(0)}$ be a starting point. Given $x^{(k)}$, the steps for obtaining $x^{(k+1)}$ are:

Step 1: If $\nabla g(x^{(k)}) = 0$, then stop. Otherwise, choose any matrix $V_k \in \partial_B G(x^{(k)})$ and let $A_k = V_k^T V_k$.

Step 2: Let $\lambda_k = g(x^{(k)})$.

Step 3: Find $d^{(k)}$ that is a solution of the linear system

$$(A_k + \lambda_k I) d^{(k)} = \nabla g(x^{(k)}).$$

Step 4: Compute the smallest nonnegative integer m_k such that

$$g(x^{(k)} + \beta^m d^{(k)}) - g(x^{(k)}) \leq -\delta \beta^m \nabla g(x^{(k)})^T d^{(k)}$$

and set

$$x^{(k+1)} = x^{(k)} + \beta^{m_k} d^{(k)}.$$

Remark 3. *(i) In Step 2, letting $\lambda_k = g(x^{(k)})$ is one of the simplest strategy because then $\{\lambda_k\}$ converges to 0.*
(ii) The line search step (Step 4) in the algorithm follows the Armijo rule.

Theorem 2. *Let $x^{(0)}$ be a starting point and $\{x^{(k)}\}$ be a sequence generated by Algorithm 1. Assume that:*

(a) $\sup_k \|V_k\| < \infty$ for all $V_k \in \partial_B G(x^{(k)})$;
(b) $\nabla g(x)$ is Lipschitzian with a constant $L_g > 0$ on the level set $L = \{x : g(x) \leq g(x^{(0)})\}$.

Then, the generalized damped Gauss–Newton method described by Algorithm 1 is well defined and either $\{x^{(k)}\}$ terminates at a stationary point of g, or else every accumulation point of $\{x^{(k)}\}$, if it exists, is a stationary point of g.

Proof. The proof is almost the same as Theorem 2.1 in [10], providing appropriately modified assumptions. □

For the nonsmooth case, the alternative condition may be considered instead of Lipschitz continuity of $\nabla g(x)$ (similar as in [10]). Thus, we have the following convergence theorem:

Theorem 3. *Let $x^{(0)}$ be a starting point and $\{x^{(k)}\}$ be a sequence generated by Algorithm 1. Assume that:*
(a) the level set $L = \left\{x : g(x) \leq g(x^{(0)})\right\}$ is bounded;
(b) G is semismooth on L.

Then, the generalized damped Gauss–Newton method described by Algorithm 1 is well defined and either $\{x^{(k)}\}$ terminates at a stationary point of g, or else every accumulation point of $\{x^{(k)}\}$, if it exists, is a stationary point of g.

Now, we take up the rate of convergence of the considered algorithm. The following theorem shows suitable conditions in various cases.

Theorem 4. *Suppose that x^* is a solution of problem (1), G is semismooth, and G is BD-regular at x^*. Then, there exists a neighborhood N_* of x^* such that, if $x^{(0)} \in N_*$ and the sequence $\{x^{(k)}\}$ is generated by Algorithm 1, we have:*
(i) $x^{(k)} \in N_*$ for all k and the sequence $\{x^{(k)}\}$ is linear convergent to x^*;
(ii) if $\delta < 0.5$, then the convergence is at least superlinear;
(iii) If G is strongly semismooth, then the convergence is quadratic.

Proof. The proof of similar theorem given by Subramanian and Xiu [10] is based on three lemmas, which have the same assumptions as theorem. Now, we present these lemmas in versions adapted to our nonsmooth case. □

Lemma 3. *Assume that d_x is a solution of the equation*

$$(A_x + \lambda_x I)d_x = \nabla g(x),$$

where

$$\lambda_x = g(x) \text{ and } A_x = V_x^T V_x$$

for some matrix V_x taken from $\partial_B G(x)$. Then, there is a neighborhood D_1 of x^* such that, for all $x \in D_1$,

$$\|x - d_x - x^*\| = o(\|x - x^*\|).$$

Lemma 4. *There is a neighborhood D_2 of x^* such that, for all $x \in D_2$,*
(a) $g(x) = \frac{1}{2}(x-x^*)^T A_*(x-x^*) + o\left(\|x-x^*\|^2\right)$,
(b) $g(x) = \frac{1}{2}(x-x^*)^T A_x(x-x^*) + o\left(\|x-x^*\|^2\right)$.

Lemma 5. *Suppose that the conditions of Lemma 1 hold. Then, there is a neighborhood D_3 of x^* such that, for all $x \in D_3$,*

$$g(x - d_x) - g(x) + \frac{1}{2}\nabla g(x)^T d_x \leq o\left(\|x - x^*\|^2\right).$$

The proofs of Lemmas 5 and 6 are almost the same as in [10]; however, in the proof of Lemma 4, we have to take into account the semismoothness and to use Lemma 1 to obtain the desired result.

At the same time, in a similar way, we may show a suitable rate of convergence.

Now, we consider the inexact version of the considered method, which computes an approximate step, using the nonnegative sequence of forcing terms to control the level of accuracy.

For the above inexact version of the algorithm, we can state the analogous theorems which are equivalents of Theorems 2–4. Based on our previous results, the proof can be carried out almost in the same way as that of theorems for the 'exact' version of the method. However, the condition (7), implied by inexactness given in Step 3 of Algorithm 2, has to be considered. Thus, we omit both theorems as proofs here.

Algorithm 2: The inexact version of the damped Gauss-Newton method for solving NCP

Let $\beta, \delta, \theta \in (0,1)$ and $\eta_k \in [0,1)$ for all k given. Let $x^{(0)} \in R^n$ be a starting point. Given $x^{(k)}$, the steps for obtaining $x^{(k+1)}$ are:

Step 1: If $\nabla g(x^{(k)}) = 0$, then stop. Otherwise, choose any matrix $V_k \in \partial_B G(x^{(k)})$ and let $A_k = V_k^T V_k$.

Step 2: Let $\lambda_k = g(x^{(k)})$.

Step 3: Find $d^{(k)}$ that is a solution of the linear system

$$\left\| (A_k + \lambda_k I) d^{(k)} + \nabla g(x^{(k)}) \right\| \leq \eta_k \left\| \nabla g(x^{(k)}) \right\|. \tag{7}$$

Step 4: If

$$\left\| \nabla g(x^{(k)} + d^{(k)}) \right\| \leq \theta \left\| \nabla g(x^{(k)}) \right\|$$

then let

$$x^{(k+1)} = x^{(k)} + d^{(k)}$$

and go to Step 1.

Step 5: Compute the smallest nonnegative integer m_k such that

$$g(x^{(k)} + \beta^m d^{(k)}) - g(x^{(k)}) \leq -\delta \beta^m \nabla g(x^{(k)})^T d^{(k)}$$

and set

$$x^{(k+1)} = x^{(k)} + \beta^{m_k} d^{(k)}$$

and go to Step 1.

---

3. Numerical Results

In this section, we present results of our numerical experiments, obtained by coding both algorithms in Code:Blocks. We use double precision on an Intel Core i7 3.2 GHz running under the Windows Server 2016 operating system. We applied the generalized damped Gauss–Newton method to solve three nonlinear complementarity problems. In the following examples: N_1 and N_2 denote the number of performed iterations to satisfy the stopping criterion $\left| x^{(k+1)} - x^{(k)} \right| < 10^{-7}$, using Algorithms 1 and 2, respectively. The forcing terms in Algorithm 2 were chosen as follows: $\eta_k = (10k)^{-1}$ for all k.

Example 1 (from Kojima and Shindo [20]). *Let the function* $F : R^4 \to R^4$ *have the form*

$$\begin{aligned}
F^1(x) &= 3x_1^2 + 2x_1 x_2 + 2x_2^2 + x_3 + 3x_4 - 6, \\
F^2(x) &= 2x_1^2 + x_1 + x_2^2 + 10x_3 + 2x_4 - 2, \\
F^3(x) &= 3x_1^2 + x_1 x_2 + 2x_2^2 + 2x_3 + 9x_4 - 9, \\
F^4(x) &= x_1^2 + 3x_2^2 + 2x_3 + 3x_4 - 3.
\end{aligned}$$

Problem (NCP) with the above function F has two solutions:

$$x^* = (1,0,3,0)^T \text{ and } x^{**} = (\sqrt{6}/2, 0, 0, 0.5)^T$$

for which

$$F(x^*) = (0, 31, 0, 4)^T \text{ and } F(x^{**}) = \left(0, 2 + \frac{\sqrt{6}}{2}, 0, 0\right)^T.$$

Thus, x^* is a non-degenerate solution of (NCP) because

$$L := \left\{i : x_i^* = 0, \ F^i(x^*) = 0\right\} = \emptyset,$$

but x^{**} is a degenerate solution.

Depending upon the starting point, we obtained the convergence iteration process to both solutions (see Table 1 or Figure 1).

Table 1. Results for Example 1.

$x^{(0)}$	N_1	N_2	Solution
$(1,0,0,0)^T$	9	11	x^{**}
$(0,0,1,0)^T$	failed	18	x^{**}
$(0,0,0,1)^T$	failed	failed	-
$(1,0,1,0)^T$	7	10	x^*
$(1,0,0,1)^T$	7	9	x^{**}
$(1,0,1,-5)^T$	6	8	x^{**}

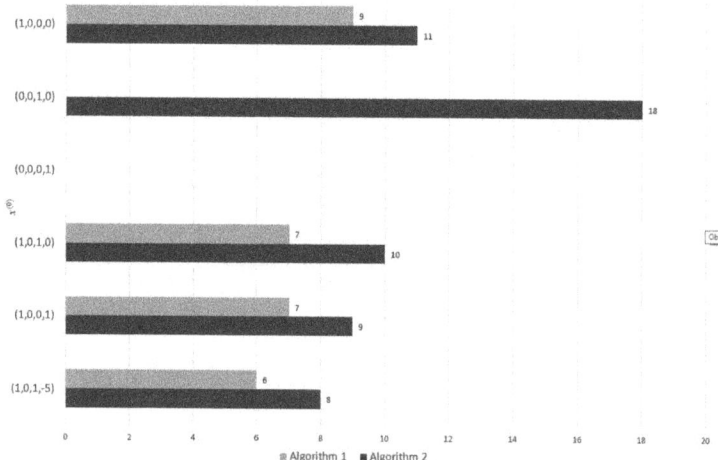

Figure 1. Number of iterations for various starting points (for Example 1).

Example 2. Let function $F : R^2 \to R^2$ be defined as follows:

$$F(x) = \begin{bmatrix} 2x_1 + x_2^2 - 6 \\ -x_1^2 + 4x_1 + \frac{1}{2}x_2 - 3 \end{bmatrix}.$$

Then, problem (NCP) has two solutions:

- non-degenerate
$$x^* = (0,6)^T \text{ for which } F(x^*) = (30,0)^T$$

- degenerate
$$x^{**} = (3,0)^T \text{ for which } F(x^{**}) = (0,0)^T.$$

Similar to Example 1, we obtained the convergence iteration process for both solutions, depending on the starting point (see Table 2 or Figure 2).

Table 2. Results for Example 2.

$x^{(0)}$	N_1	N_2	Solution
$(0,0)^T$	5	7	x^*
$(1,0)^T$	1	2	x^{**}
$(0,1)^T$	4	7	x^*
$(1,-1)^T$	2	4	x^{**}
$(-1,1)^T$	4	7	x^{**}
$(5,5)^T$	3	5	x^*
$(100,100)^T$	3	6	x^*

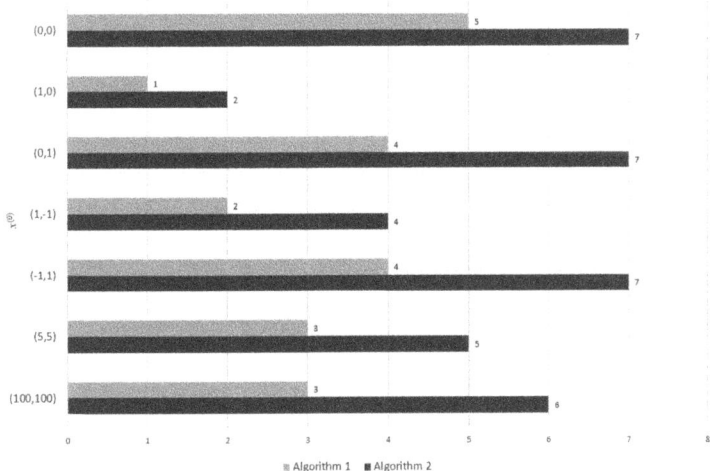

Figure 2. Number of iterations for various starting points (for Example 2.)

Example 3 (from Jiang and Qi [21]). Let function $F: R^n \to R^n$ has the form $F(x) = Mx + q$, where

$$M = \begin{bmatrix} 4 & -1 & 0 & \ldots & 0 & 0 \\ -1 & 4 & -1 & \ldots & 0 & 0 \\ 0 & -1 & 4 & \ldots & 0 & 0 \\ \ldots & \ldots & \ldots & \ldots & \ldots & \ldots \\ 0 & 0 & 0 & \ldots & 4 & -1 \\ 0 & 0 & 0 & \ldots & -1 & 4 \end{bmatrix}, \quad q = (-1, \ldots, -1)^T.$$

Because F is strictly monotonic, the proper problem (NCP) has exactly one solution.

Calculations have been made for various n with one starting point $x^{(0)} = (0, \ldots, 0)^T$. For all tests, we obtain the same number of iterations $N_1 = 3$ and $N_2 = 4$.

4. Conclusions

We have given the nonsmooth version of the damped generalized Gauss–Newton method presented by Subramanian and Xu [10]. The generalized Newton algorithms related to the Gauss–Newton method are well-known important tools for solving nonsmooth equations, which arise from various nonlinear problems such as nonlinear complementarity or variational inequality. These algorithms are especially useful when the problem has many variables. We have proved that the sequences generated by the methods are superlinearly convergent under mild assumptions. Clearly, the semismoothness and BD-regularity are sufficient to obtain only a superlinear convergence of methods, while strong semismoothness even gives quadratic convergence. However, if function G is not semismooth or BD-regular or the gradient of g is not Lipschitzian, the Gauss–Newton methods may be useless.

The performance of both methods was evaluated in terms of the number of iterations required. The analysis of the numerical results seems to indicate that the methods are usually reliable for solving semismooth problems. The results show that the inexact approach can produce a noticeable slowdown by the number of iterations (compare N_1 and N_2 in Figures 1 and 2). In turn, an important advantage is that the algorithms allow us to find various solutions to the problem (this can be observed in two examples: the first and second one). However, if there are many solutions of the problem, then the relationship between the starting point and the obtained solution may be unpredictable.

Clearly, traditional numerical algorithms aren't the only method for solving the nonlinear complementarity problems, regardless of the degree of nonsmoothness. Except for the methods presented in the paper and mentioned in the Introduction, some computational intelligence algorithms can be used to solve (NCP) problems, i.a., monarch butterfly optimization (see [22,23]), the earthworm optimization algorithm (see [24]), the elephant herding optimization (see [25,26]), or the moth search algorithm (see [27,28]). All of these approaches are bio-inspired metaheuristic algorithms.

Funding: This research received no external funding.

Conflicts of Interest: The author declares no conflict of interest.

References

1. Mangasarian, O.L. Equivalence of the complementarity problem to a system of nonlinear equations. *SIAM J. Appl. Math.* **1976**, *31*, 89–92. [CrossRef]
2. Harker, P.T.; Pang, J.S. Finite-dimensional variational inequality and nonlinear complementarity problems: A survey of theory, algorithms and applications. *Math. Program.* **1990**, *48*, 161–220. [CrossRef]
3. Ferris, M.C.; Pang, J.S. Engineering and economic applications of complementarity problems. *SIAM Rev.* **1997**, *39*, 669–713. [CrossRef]
4. Ortega, J.M.; Rheinboldt, W.C. *Iterative Solution of Nonlinear Equations in Several Variables*; Academic Press: New York, NY, USA, 1970.
5. Chen, J.; Li, W. Local convergence of Gauss–Newton's like method in weak conditions. *J. Math. Anal. Appl.* **2006**, *324*, 381–394. [CrossRef]
6. Fan, J.; Pan, J. On the convergence rate of the inexact Levenberg-Marquardt method. *J. Ind. Manag. Optim.* **2011**, *7*, 199–210. [CrossRef]
7. Yamashita, N.; Fukushima, M. On the Rate of Convergence of the Levenberg-Marquardt Method. *J. Comput. Suppl.* **2001**, *15*, 227–238.
8. De Luca, T.; Facchinei, F.; Kanzow, C.T. A theoretical and numerical comparison of some semismooth algorithms for complementarity problems. *Comp. Optim. Appl.* **2000**, *16*, 173–205. [CrossRef]
9. Xiu, N.; Zhang, J. A smoothing Gauss–Newton method for the generalized HLCP. *J. Comput. Appl. Math.* **2001**, *129*, 195–208. [CrossRef]
10. Subramanian, P.K.; Xiu, N.H. Convergence analysis of Gauss–Newton method for the complemetarity problem. *J. Optim. Theory Appl.* **1997**, *94*, 727–738. [CrossRef]
11. Martínez, J.M.; Qi, L. Inexact Newton method for solving nonsmooth equations. *J. Comput. Appl. Math.* **1995** *60*, 127–145. [CrossRef]

12. Bao, J.F.; Li, C.; Shen, W.P.; Yao, J.C.; Guu, S.M. Approximate Gauss–Newton methods for solving underdetermined nonlinear least squares problems. *App. Num. Math.* **2017**, *111*, 92–110. [CrossRef]
13. Cartis, C.; Roberts L. A derivative-free Gauss–Newton method. *Math. Program. Comput.* **2019**, *11*, 631–674. [CrossRef]
14. Clarke, F.H. *Optimization and Nonsmooth Analysis*; John Wiley & Sons: New York, NY, USA, 1983.
15. Qi, L. Convergence analysis of some algorithms for solving nonsmooth equations. *Math. Oper. Res.* **1993**, *18*, 227–244. [CrossRef]
16. Mifflin, R. Semismooth and semiconvex functions in constrained optimization. *SIAM J. Control Optim.* **1977**, *15*, 142–149. [CrossRef]
17. Qi, L.; Sun, D. A nonsmooth version of Newton's method. *Math. Program.* **1993**, *58*, 353–367. [CrossRef]
18. Pang, J.S.; Qi, L. Nonsmooth equations: Motivation and algorithms. *SIAM J. Optim.* **1993**, *3*, 443–465. [CrossRef]
19. Ulbrich, M. Nonmonotone trust-region methods for bound-constrained semismooth systems of equations with applications to nonlinear mixed complementarity problems. *SIAM J. Optim.* **2001**, *11*, 889–917. [CrossRef]
20. Kojima, M.; Shindo, S. Extensions of Newton and quasi-Newton methods to systems of PC^1-equations. *J. Oper. Res. Soc. Jpn.* **1986**, *29*, 352–374. [CrossRef]
21. Jiang, H.; Qi, L. A new nonsmooth equations approach to nonlinear complementarity problems. *SIAM J. Control Optim.* **1997**, *35*, 178–193. [CrossRef]
22. Feng, Y.; Wang, G.; Li, W.; Li, N. Multi-strategy monarch butterfly optimization algorithm for discounted 0–1 knapsack problem. *Neural Comput. Appl.* **2018**, *30*, 3019–3036. [CrossRef]
23. Feng, Y.; Yu, X.; Wang, G. A novel monarch butterfly optimization with global position updating operator for large-scale 0–1 knapsack problems. *Mathematics* **2019**, *7*, 1056. [CrossRef]
24. Wang, G.; Deb, S.; Coelho, L.D. Earthworm optimisation algorithm: A bio-inspired metaheuristic algorithm for global optimisation problems. *Int. J. Biol. Inspired Comput.* **2018**, *12*, 1–22. [CrossRef]
25. Wang, G.; Deb, S.; Coelho, L.D. Elephant herding optimization. In Proceedings of the 2015 3rd International Symposium on Computational and Business Intelligence (ISCBI), Bali, Indonesia, 7–9 December 2015; pp. 1–5.
26. Wang, G.; Deb, S.; Gao, X.; Coelho, L.D. A new metaheuristic optimisation algorithm motivated by elephant herding behaviour. *Int. J. Biol. Inspired Comput.* **2016**, *8*, 394–409. [CrossRef]
27. Feng, Y.; Wang, G. Binary moth search algorithm for discounted 0-1 knapsack problem. *IEEE Access* **2018**, *6*, 10708–10719. [CrossRef]
28. Wang, G. Moth search algorithm: A bio-inspired metaheuristic algorithm for global optimization problems. *Memet. Comput.* **2018**, *10*, 151–164. [CrossRef]

 © 2020 by the author. Licensee MDPI, Basel, Switzerland. This article is an open access article distributed under the terms and conditions of the Creative Commons Attribution (CC BY) license (http://creativecommons.org/licenses/by/4.0/).

Article

Polyhedral DC Decomposition and DCA Optimization of Piecewise Linear Functions

Andreas Griewank and Andrea Walther *

Institut für Mathematik, Humboldt-Universität zu Berlin, 10099 Berlin, Germany; griewank@math.hu-berlin.de
* Correspondence: andrea.walther@math.hu-berlin.de

Received: 28 May 2020; Accepted: 8 July 2020; Published: 11 July 2020

Abstract: For piecewise linear functions $f : \mathbb{R}^n \mapsto \mathbb{R}$ we show how their abs-linear representation can be extended to yield simultaneously their decomposition into a convex $\check{f}$ and a concave part $\hat{f}$, including a pair of generalized gradients $\check{g} \in \mathbb{R}^n \ni \hat{g}$. The latter satisfy strict chain rules and can be computed in the reverse mode of algorithmic differentiation, at a small multiple of the cost of evaluating f itself. It is shown how $\check{f}$ and $\hat{f}$ can be expressed as a single maximum and a single minimum of affine functions, respectively. The two subgradients $\check{g}$ and $-\hat{g}$ are then used to drive DCA algorithms, where the (convex) inner problem can be solved in finitely many steps, e.g., by a Simplex variant or the true steepest descent method. Using a reflection technique to update the gradients of the concave part, one can ensure finite convergence to a local minimizer of f, provided the Linear Independence Kink Qualification holds. For piecewise smooth objectives the approach can be used as an inner method for successive piecewise linearization.

Keywords: DC function; abs-linearization; DCA

---

1. Introduction and Notation

There is a large class of functions $f : \mathbb{R}^n \mapsto \mathbb{R}$ that are called DC because they can be represented as the difference of two convex functions, see for example [1,2]. This property can be exploited in various ways, especially for (hopefully global) optimization. We find it notationally and conceptually more convenient to express these functions as averages of a convex and a concave function such that

$$f(x) = \tfrac{1}{2}(\check{f}(x) + \hat{f}(x)) \quad \text{with} \quad \check{f}(x) \text{ convex and } \hat{f}(x) \text{ concave.}$$

Throughout we will annotate the convex part by superscript ˇ and the concave part by superscript ˆ, which seems rather intuitive since they remind us of the absolute value function and its negative. Since we are mainly interested in piecewise linear functions we assume without much loss of generality that the functions f and the convex and concave components are well defined and finite on all of the Euclidean space $\mathbb{R}^n$. Allowing both components to be infinite outside their proper domain would obviously generate serious indeterminacies, i.e., NaNs in the numerical sense. As we will see later we can in fact ensure in our setting that pointwise

$$\hat{f}(x) \leq f(x) \leq \check{f}(x) \quad \text{for all} \quad x \in \mathbb{R}^n, \tag{1}$$

which means that we actually obtain an inclusion in the sense of interval mathematics [3]. This is one of the attractions of the averaging notation. We will therefore also refer to $\hat{f}$ and $\check{f}$ as the concave and convex bounds of f.

Conditioning of the Decomposition

In parts of the literature the two convex functions $\check{f}$ and $-\hat{f}$ are assumed to be nonnegative, which has some theoretical advantages. In particular, see, e.g., [4], one obtains for the square $h = f^2$ of a DC function f the decomposition

$$h = \tfrac{1}{4}(\check{f}+\hat{f})^2 = \tfrac{1}{2}\{\underbrace{\tfrac{1}{4}(\check{f}^2+\hat{f}^2)}_{\equiv \check{h}} + \underbrace{\tfrac{1}{4}[-(\check{f}-\hat{f})^2]}_{\equiv \hat{h}}\}. \tag{2}$$

The sign conditions of $\check{f}$ and $\hat{f}$ are necessary to ensure that the three squares on the right hand side are convex functions. Using the Apollonius identity $f \cdot h = \tfrac{1}{2}[(f+h)^2 - f^2 - h^2]$ one may then deduce in a constructive way that not only sums but also products of DC functions inherit this property. In general, since the convex functions $\check{f}$ and $-\hat{f}$ have both supporting hyperplanes one can at least theoretically always find positive coefficients α and β such that

$$\check{f}(x) + \alpha + \beta\|x\|^2 \geq 0 \geq \hat{f}(x) - \alpha - \beta\|x\|^2 \quad \text{for} \quad x \in \mathbb{R}^n.$$

Then the average of these modified functions is still f and their respective convexity/concavity properties are maintained. In fact, this kind of proximal shift can be used to show that any twice Lipschitz continuously differentiable function is DC, which raises the suspicion that the property by itself does not provide all that much exploitable structure from a numerical point of view. We believe that for its use in practical algorithms one has to make sure or simply assume that the condition number

$$\kappa(\check{f},\hat{f}) \equiv \sup_{x\in\mathbb{R}^n} \frac{|\check{f}(x)|+|\hat{f}(x)|}{|\check{f}(x)+\hat{f}(x)|} \in [1,\infty]$$

is not too large. Otherwise, there is the danger that the value of f is effectively lost in the rounding error of evaluating $\check{f}+\hat{f}$. For sufficiently large quadratic shifts of the nature specified above one has $\kappa \sim \beta$. The danger of an excessive growth in κ seems akin to the successive widening in interval calculations and similarly stems also from the lack of strict arithmetic rules. For example doubling f and then subtracting it yields the successive decompositions

$$(2f) - f = (\check{f}+\hat{f}) - \tfrac{1}{2}(\check{f}+\hat{f}) = (\check{f} - \tfrac{1}{2}\hat{f}) + (\hat{f} - \tfrac{1}{2}\check{f}) = \tfrac{1}{2}[(2\check{f}-\hat{f}) + (2\hat{f}-\check{f})]. \tag{3}$$

If in Equation (3) by chance we had originally $-\hat{f} = \tfrac{1}{2}\check{f} > 0$ so that $f = \tfrac{1}{2}\check{f}$ with the condition number $\kappa(\check{f}, -0.5\check{f}) = 3$ we would get after the doubling and subtraction the condition number $\kappa(2.5\check{f}, -2\check{f}) = 9$. So it is obviously important that the original algorithm avoids as much as possible calculations that are ill-conditioned in that they even just partly compensate each other.

Throughout the paper we assume that the functions in question are evaluated by a computational procedure that generates a sequence of intermediate scalars, which we denote generically by u, v and w. The last one of these scalar variables is the dependent, which is usually denoted by f. All of them are continuous functions $u = u(x)$ of the vector $x \in \mathbb{R}^n$ of independent variables. As customary in mathematics we will often use the same symbol to identify a function and its dependent variable. For the overall objective we will sometimes distinguish them and write $y = f(x)$. For most of the paper we assume that the intermediates are obtained from each other by affine operations or the absolute value function so that the resulting $u(x)$ are all piecewise linear functions.

The paper is organized as follows. In the following Section 2 we develop rules for propagating the convex/concave decomposition through a sequence of abs-linear operations applied to intermediate quantities u. This can be done either directly on the pair of bounds $(\check{u}, \hat{u})$ or on their average u and their halved distance $\delta u = \tfrac{1}{2}(\check{u} - \hat{u})$. In Section 3 we organize such sequences into an abs-linear form for f and then extend it to simultaneously yield the convex/concave decomposition. As a consequence of this analysis we get a strengthened version of the classical max−min representation of piecewise linear functions, which reduces to the difference of two polyhedral parts in max- and min-form. In Section 4 we develop strict rules for propagating certain generalized gradient pairs $(\check{g}, \hat{g})$ of $(\check{u}, \hat{u})$ exploiting convexity and the cheap gradient principle [5]. In Section 5 we discuss the consequences for the DCA when using limiting gradients $(\check{g}, \hat{g})$, solving the inner, linear optimization problem (LOP) exactly, and ensuring optimality via polyhedral reflection. In Section 6 we demonstrate the new results on the nonconvex and piecewise linear chained Rosenbrock version of Nesterov [6]. Section 7 contains a summary and preliminary conclusion with outlook. In the Appendix A we give the details of the necessary and sufficient optimality test from [7] in the present DC context.

2. Propagating Bounds and/or Radii

In Equation (3) we already assumed that doubling is done componentwise and that for a difference $v = w - u$ of DC functions w and u, one defines the convex and concave parts by

$$\widetilde{(w-u)} = \widetilde{w} - \hat{u} \quad \text{and} \quad \widehat{(w-u)} = \hat{w} - \check{u},$$

respectively. This yields in particular for the negation

$$\widetilde{(-u)} = -\hat{u} \quad \text{and} \quad \widehat{(-u)} = -\check{u}. \tag{4}$$

For piecewise linear functions we need neither the square formula Equation (2) nor the more general decompositions for products. Therefore we will not insist on the sign conditions even though they would be also maintained automatically by Equation (4) as well as the natural linear rules for the convex and concave parts of the sum and the multiple of a DC function, namely

$$\widetilde{(w+u)} = (\check{w}+\check{u}) \quad \text{and} \quad \widehat{(w+u)} = (\hat{w} + \hat{u}),$$
$$\widetilde{(cu)} = c\check{u} \quad \text{and} \quad \widehat{(cu)} = c\hat{u} \quad \text{if} \quad c \geqslant 0,$$
$$\widetilde{(cu)} = c\hat{u} \quad \text{and} \quad \widehat{(cu)} = c\check{u} \quad \text{if} \quad c \leqslant 0.$$

However, the sign conditions would force one to decompose simple affine functions $u(x) = a^\top x + \beta$ as

$$u(x) = \max(0, a^\top x + \beta) + \min(0, a^\top x + \beta) \equiv \tfrac{1}{2}(\check{u}(x) + \hat{u}(x)), \tag{5}$$

which does not seem such a good idea from a computational point of view.

The key observation for this paper is that as is well known (see e.g., [8]), one can propagate the absolute value operation according to the identity

$$|u| = \max(u, -u) = \tfrac{1}{2}\max(\check{u} + \hat{u}, -\check{u} - \hat{u})$$
$$= \max(\check{u}, -\hat{u}) + \tfrac{1}{2}(\hat{u} - \check{u})$$
$$\iff \widetilde{|u|} = 2\max(\check{u}, -\hat{u}) \quad \text{and} \quad \widehat{|u|} = \hat{u} - \check{u}. \tag{6}$$

Here the equality in the second line can be verified by shifting the difference $\tfrac{1}{2}(\hat{u} - \check{u})$ into the two arguments of the max. Again we see that when applying the absolute value operation to an already positive convex function $u = \tfrac{1}{2}\check{u} \geqslant 0$ we get $\widetilde{|u|} = 2\check{u}$ and $\widehat{|u|} = -\check{u}$ so that the condition number grows from $\kappa(\check{u}, 0) = 1$ to $\kappa(2\check{u}, -\check{u}) = 3$. In other words, we observe once more the danger that both component functions drift apart. This looks a bit like simultaneous growth of numerator and denominator in rational arithmetic, which can sometimes be limited through cancelations by common integer factors. It is currently not clear when and how a similar compactification of a given convex/concave decomposition can be achieved. The corresponding rule for the maximum is similarly easy derived, namely

$$\max(u,w) = \tfrac{1}{2}\max(\check{u}+\hat{u},\check{w}+\hat{w}) = \tfrac{1}{2}\left(\max(\check{u}-\hat{w},\check{w}-\hat{u}) + (\hat{u}+\hat{w})\right).$$

When u and w as well as their decomposition are identical we arrive at the new decomposition $u = \max(u,u) = \tfrac{1}{2}((\check{u}-\hat{u}) + 2\hat{u})$, which obviously represents again some deterioration in the conditioning.

While it was pointed out in [4] that the DC functions $u = \tfrac{1}{2}(\check{u}+\hat{u})$ themselves form an algebra, their decomposition pairs $(\check{u},\hat{u})$ are not even an additive group, as only the zero $(0,0)$ has a negative partner, i.e., an additive inverse. Naturally, the pairs $(\check{u},\hat{u})$ form the Cartesian product between the convex cone of convex functions and its negative, i.e., the cone of concave functions. The DC functions are then the linear envelope of the two cones in some suitable space of locally Lipschitz continuous functions. It is not clear whether this interpretation helps in some way, and in any case we are here mainly concerned with piecewise linear functions.

Propagating the Center and Radius

Rather than propagating the pairs $(\check{u},\hat{u})$ through an evaluation procedure as defined in [5] to calculate the function value $f(x)$ at a given point x, it might be simpler and better for numerical stability to propagate the pair

$$u = \tfrac{1}{2}(\check{u}+\hat{u}) \;\wedge\; \delta u = \tfrac{1}{2}(\check{u}-\hat{u}) \iff \check{u} = u + \delta u \;\wedge\; \hat{u} = u - \delta u. \tag{7}$$

This representation resembles the so-called central form in interval arithmetic [9] and we will call therefore u the central value and δu the radius. In other words, u is just the normal piecewise affine intermediate function and the δu is a convex distance function to the hopefully close convex and concave part. Should the potential blow-up discussed above actually occur, this will only effect δu but not the central value u itself. Moreover, at least theoretically one might decide to reduce δu from time to time making sure of course that the corresponding $\check{u}$ and $\hat{u}$ as defined in Equation (7) stay convex and concave, respectively. The condition number now satisfies the bound

$$\kappa(u+\delta u, u-\delta u) = \sup_x \frac{|u+\delta u| + |u-\delta u|}{2|u|}$$
$$= \sup_x \tfrac{1}{2}\left\{\left|1+\frac{\delta u}{u}\right| + \left|1-\frac{\delta u}{u}\right|\right\} \leq 1 + \sup_x \left|\frac{\delta u}{u}\right|.$$

Recall here that all intermediate quantities $u = u(x)$ are functions of the independent variable vector $x \in \mathbb{R}^n$. Naturally, we will normally only evaluate the intermediate pairs u and δu at a few iterates of whatever numerical calculation one performs involving f so that we can only sample the ratio

$$\rho u(x) \equiv |\delta u(x)/u(x)|$$

pointwise, where the denominator is hopefully nonzero. We will also refer to this ratio as the relative gap of the convex/concave decomposition at a certain evaluation point x. The arithmetic rules for propagating radii of the central form in central convex/concave arithmetic are quite simple.

Lemma 1 (Propagation rules for central form). *With $c, d, x \in \mathbb{R}$ two constants and an independent variable we have*

$$\begin{aligned}
v = c + dx &\implies \delta v = 0 &&\implies \rho v = 0 \text{ if } v \neq 0 \\
v = u \pm w &\implies \delta v = \delta u + \delta w &&\implies \rho v \leq \tfrac{|u|+|w|}{|u \pm w|}\max(\rho u, \rho w) \\
v = cu &\implies \delta v = |c|\,\delta u &&\implies \rho v = \rho u \text{ if } c \neq 0 \\
v = |u| &\implies \delta v = |u| + 2\delta u &&\implies \rho v \in [1, 1+2\rho u].
\end{aligned} \tag{8}$$

Proof. The last rule follows from Equation (6) by

$$\begin{aligned}
\delta(|u|) &= \tfrac{1}{2}\left(\widetilde{|u|} - \widehat{|u|}\right) = \max(\check{u}, -\hat{u}) - \tfrac{1}{2}(\hat{u} - \check{u}) \\
&= \max(\check{u} - \delta u, -\hat{u} - \delta u) + 2\,\delta u \\
&= \max(u, -u) + 2\,\delta u = |u| + 2\,\delta u\,.
\end{aligned}$$

□

The first equation in Equation (8) means that for all quantities u that are affine functions of the independent variables x the corresponding radius δu is zero so that $\check{u} = u = \hat{u}$ until we reach the first absolute value. Notice that δv does indeed grow additively for the subtraction just like for the addition. By induction it follows from the rules above for an inner product that

$$\delta\left(\sum_{j=1}^{m} c_j u_j\right) = \sum_{j=1}^{m} |c_j|\,\delta u_j\,, \qquad (9)$$

where the $c_j \in \mathbb{R}$ are assumed to be constants. As we can see from the bounds in Lemma 1 the relative gap can grow substantially whenever one performs an addition of values with opposite sign or applies the absolute value operation. In contrast to interval arithmetic on smooth functions one sees that the relative gap, though it may be zero or small initially immediately jumps above 1 when one hits the first absolute value operation. This is not really surprising since the best concave lower bound on $u(x) = |x|$ itself is $\hat{u}(x) = 0$ so that $\delta u = |x|$, $\check{u}(x) = 2|x|$ and thus $\rho u(x) = 1$ constantly. On the positive side one should notice that throughout we do not lose sight of the actual central values $u(x)$, which can be evaluated with full arithmetic precision. In any case we can think of neither ρ nor $\kappa \leqslant 1 + \rho$ as small numbers, but we must be content if they do not actually explode too rapidly. Therefore they will be monitored throughout our numerical experiments.

Again we see that the computational effort is almost exactly doubled. The radii can be treated as additional variables that occur only in linear operations and stay nonnegative throughout. Notice that in contrast to the (nonlinear) interval case we do not loose any accuracy by propagating the central form. It follows immediately by induction from Lemma 1 that any function evaluated by a evaluation procedure that comprises a finite sequence of

- initializations to independent variables
- multiplications by constants
- additions or subtractions
- absolute value applications

is piecewise affine and continuous. We will call these operations and the resulting evaluation procedure abs-linear. It is also easy to see that the absolute values $|\cdot|$ can be replaced by the maximum $\max(\cdot, \cdot)$ or the minimum $\min(\cdot, \cdot)$ or the positive part function $\max(0, \cdot)$ or any combination of them, since they can all be mutually expressed in terms of each other and some affine operations. Conversely, it follows from the min-max representation established in [10] (Proposition 2.2.2) that any piecewise affine function f can be evaluated by such an evaluation procedure. Consequently, by applying the formulas Equations (4)–(6) one can propagate at the same time the convex and concave components for all intermediate quantities. Alternatively, one can propagate the centered form according to the rules given in Lemma 1. These rules are also piecewise affine so that we have a finite procedure for simultaneously evaluating $\check{u}$ and $\hat{u}$ or u and δu as piecewise linear functions. The combined computation requires about 2–3 times as many arithmetic operations and twice as many memory accesses. Of course due to the interdependence of the two components it is not possible to evaluate just one of them without the other. As we will see the same is true for the generalized gradients to be discussed later in Section 4.

3. Forming and Extending the Abs-Linear Form

In practice all piecewise linear objectives can be evaluated by a sequence of abs-linear operations, possibly after min and max have been rewritten as

$$\min(u,w) = \tfrac{1}{2}(u+w-|u-w|) \quad \text{and} \quad \max(u,w) = \tfrac{1}{2}(u+w+|u-w|) \,. \tag{10}$$

Our only restriction is that the number s of intermediate scalar quantities, say z_i, is fixed, which is true for example in the max − min representation. Then we can immediately cast the procedure in matrix-vector notation as follows:

Lemma 2 (Abs-Linear Form). *Any continuous piecewise affine function $f : x \in \mathbb{R}^n \mapsto y \in \mathbb{R}$ can be represented by*

$$\begin{aligned} z &= c + Zx + Mz + L|z| \,, \\ y &= d + a^\top x + b^\top z \,, \end{aligned} \tag{11}$$

where $z \in \mathbb{R}^s, Z \in \mathbb{R}^{s \times n}, M, L \in \mathbb{R}^{s \times s}$ strictly lower triangular, $d \in \mathbb{R}, a \in \mathbb{R}^n, b \in \mathbb{R}^s$ and $|z|$ denotes the componentwise modulus of the vector z.

It should be noted that the construction of this general abs-linear form requires no analysis or computation whatsoever. However, especially for our purpose of generating a reasonably tight DC decomposition, it is advantages to reduce the size of the abs-normal form by eliminating all intermediates z_j with $j < s$ for which $|z_j|$ never occurs on the right hand side. To this end we may simply substitute the expression of z_j given in the j-th row in all places where z_j itself occurs on the right hand side. The result is what we will call a reduced abs-normal form, where after renumbering, all remaining z_j with $j < s$ are switching variables in that $|z_j|$ occurs somewhere on the right hand side. In other words, all but the last column of the reduced, strictly lower triangular matrix L are nontrivial. Again, this reduction process is completely mechanical and does not require any nontrivial analysis, other than looking up which columns of the original L were zero. The resulting reduced system is smaller and probably denser, which might increase the computation effort for evaluating f itself. However, in view of Equation (9) we must expect that for the reduced form the radii will grow slower if we first accumulate linear coefficients and then take their absolute values. Hence we will assume in the remainder of this paper that the abs-normal form for our objective f of interest is reduced.

Based on the concept of abs-linearization introduced in [11], a slightly different version of a (reduced) abs-normal form was already proposed in [12]. Now in the present paper, both z and y depend directly on z via the matrix M and the vector b, but y does no longer depend directly on $|z|$. All forms can be easily transformed into each other by elementary modifications. The intermediate variables z_i can be calculated successively for $1 \leq i \leq s$ by

$$z_i = c_i + Z_i x + M_i z + L_i |z| \,, \tag{12}$$

where Z_i, M_i and L_i denote the ith rows of the corresponding matrix. By induction on i one sees immediately that they are piecewise affine functions $z_i = z_i(x)$, and we may define for each x the signature vector

$$\sigma(x) = (\text{sgn}(z_i(x)))_{i=1...s} \in \{-1, 0, 1\}^s \,.$$

Consequently we get the inverse images

$$\mathcal{P}_\sigma \equiv \{x \in \mathbb{R}^n : \text{sgn}(z(x)) = \sigma\} \quad \text{for} \quad \sigma \in \{-1, 0, 1\}^s \,, \tag{13}$$

which are relatively open polyhedra that form collectively a disjoint decomposition of $\mathbb{R}^n$. The situation for the second example of Nesterov is depicted in Figure 3 in the penultimate section. There are six polyhedra of full dimension, seven polyhedra of co dimension 1 drawn in blue and two points, which are polyhedra of dimension 0. The point $(0, -1)$ with signature $(0, -1, 0)$ is stationary and the point $(1, 1)$ with signature $(1, 0, 0)$ is the minimizer as shown in [7]. The arrows indicate the path of our reflection version of the DCA method as described in Section 5.

When σ is definite, i.e., has no zero components, which we will denote by $0 \notin \sigma$, it follows from the continuity of $z(x)$ that $\mathcal{P}_\sigma$ has full dimension n unless it is empty. In degenerate situations this may also be true for indefinite σ but then the closure of $\mathcal{P}_\sigma$ is equal to the extended closure

$$\bar{\mathcal{P}}_{\tilde{\sigma}} \equiv \{x \in \mathbb{R}^n : \sigma(x) < \tilde{\sigma}\} \supset \text{close}(\mathcal{P}_{\tilde{\sigma}}) \tag{14}$$

for some definite $0 \notin \tilde{\sigma} > \sigma$. Here the (reflexive) partial ordering $<$ between the signature vectors satisfies the equivalence

$$\tilde{\sigma} < \sigma \quad \Longleftrightarrow \quad \tilde{\sigma}_i \sigma_i \leq \sigma_i^2 \text{ for } i = 1 \dots s \quad \Longleftrightarrow \quad \bar{\mathcal{P}}_{\tilde{\sigma}} \subset \bar{\mathcal{P}}_\sigma$$

as shown in [13]. One can easily check that for any $\sigma > \tilde{\sigma}$ there exists a unique signature

$$(\sigma \triangleright \tilde{\sigma})_i = \begin{cases} \sigma_i & \text{if } \tilde{\sigma}_i \neq 0 \\ -\sigma_i & \text{if } \tilde{\sigma}_i = 0 \end{cases} \quad \text{for } i = 1 \dots s \tag{15}$$

We call $\tilde{\sigma} \equiv \sigma \triangleright \tilde{\sigma}$ the reflection of σ at $\tilde{\sigma}$, which satisfies also $\sigma > \tilde{\sigma}$ and we have in fact $\bar{\mathcal{P}}_\sigma \cap \bar{\mathcal{P}}_{\tilde{\sigma}} = \bar{\mathcal{P}}_{\tilde{\sigma}}$. Hence the relation between σ and $\tilde{\sigma}$ is symmetric in that also $\sigma = \tilde{\sigma} \triangleright \tilde{\sigma}$. Therefore we will call $(\sigma, \tilde{\sigma})$ a complementary pair with respect to $\tilde{\sigma}$. In the very special case $z_i = x_i$ for $i = 1 \dots n = s - 1$ the $\bar{\mathcal{P}}_\sigma$ are orthants and their reflections at the origin $\{0\} = \bar{\mathcal{P}}_0 \subset \mathbb{R}^n$ are their geometric opposites $\bar{\mathcal{P}}_{\tilde{\sigma}}$ with $\tilde{\sigma} = -\sigma$. Here one can see immediately that all edges, i.e., one-dimensional polyhedra, have Cartesian signatures $\pm e_i$ for $i = 1 \dots n$ and belong to $\bar{\mathcal{P}}_\sigma$ or $\bar{\mathcal{P}}_{\tilde{\sigma}}$ for any given σ. Notice that $\mathring{x}$ is a local minimizer of a piecewise linear function if and only if it is a local minimizer along all edges of nonsmoothness emanating form it. Consequently, optimality of f restricted to a complementary pair is equivalent to local optimality on $\mathbb{R}^n$, not only in this special case, but whenever the Linear Independence Kink Qualification (LIKQ) holds as introduced in [13] and defined in the Appendix A. This observation is the basis of the implicit optimality condition verified by our DCA variant Algorithm 1 through the use of reflections. The situation is depicted in Figure 3 where the signatures $(-1, -1, -1)$ and $(1, -1, 1)$ as well as $(1, -1, 1)$ and $(1, 1, -1)$ form complementary pairs at $(0, -1)$ and $(1, 1)$, respectively. At both reflection points there are four emanating edges, which all belong to one of the three polyhedra mentioned.

Applying the propagation rules from Lemma 1, one obtains with $\delta x = 0 \in \mathbb{R}^n$ the recursion

$$\delta z_1 = \delta(c_1 + Z_1 x) = 0$$
$$\delta z_i = (|M_i| + 2|L_i|)\delta z + |L_i||z| \quad \text{for } i = 2 \dots s,$$

where the modulus is once more applied componentwise for vectors and matrices. Hence, we have again in matrix vector notation

$$\delta z = (|M| + 2|L|)\delta z + |L||z|, \tag{16}$$

which yields for δz the explicit expression

$$\delta z = (I - |M| - 2|L|)^{-1}|L||z| = \sum_{j=0}^{v}(|M| + 2|L|)^j |L||z| \geq 0. \tag{17}$$

Here, ν is the so-called switching depth of the abs-linear form of f, namely the largest $\nu \in \mathbb{N}$ such that $(|M| + |L|)^\nu \neq 0$, which is always less than s due to the strict lower triangularity of M and L. The unit lower triangular $(I - |M| - 2|L|)$ is an M-matrix [14], and interestingly enough does not even depend on x but directly maps $|z| = |z(x)|$ to $\delta z = \delta z(x)$. For the radius of the function itself, the propagation rules from Lemma 1 then yield

$$\delta f(x) = \delta y = |b|^\top \delta z \geq 0 \,. \tag{18}$$

This nonnegativity implies the inclusion Equation (1) already mentioned in Section 1, i.e.:

Theorem 1 (Inclusion by convex/concave decomposition). *For any piecewise affine function f in abs-linear form, the construction defined in Section 2 yields a convex/concave inclusion*

$$\hat{f}(x) \leq f(x) \equiv \tfrac{1}{2}(\check{f}(x) + \hat{f}(x)) \leq \check{f}(x) \,.$$

Moreover, the convex and the concave parts $\check{f}(x)$ and $\hat{f}(x)$ have exactly the same switching structure as $f(x)$ in that they are affine on the same polyhedra $\mathcal{P}_\sigma$ defined in (13).

Proof. Equations (16) and (17) ensure that $\delta f(x)$ is nonnegative at all $x \in \mathbb{R}^n$ such that

$$\hat{f}(x) = f(x) - \delta f(x) \leq f(x) \leq f(x) + \delta f(x) \leq \check{f}(x) \,.$$

It follows from Equation (17) that the radii $\delta z_i(x)$ are like the $|z_i(x)|$ piecewise linear with the only nonsmoothness arising through the switching variables $z(x)$ themselves. Obviously this property is inherited by $\delta f(x)$ and the linear combinations $\check{f}(x) = f(x) + \delta f(x)$ and $\hat{f}(x) = f(x) - \delta f(x)$, which completes the proof. □

Combining Equations (16) and (18) with the abs-linear form of the piecewise affine function f and defining $\tilde{z} = (z, \delta z) \in \mathbb{R}^{2s}$, one obtains for the calculation of $\check{f}(x) \equiv \tilde{y} \equiv (y, \delta y)$ the following abs-linear form

$$\tilde{z} = \tilde{c} + \tilde{Z}x + \tilde{M}\tilde{z} + \tilde{L}|\tilde{z}| \,, \tag{19}$$

$$\tilde{y} = \tilde{d} + \tilde{a}^\top x + \tilde{b}^\top \tilde{z} \tag{20}$$

with the vectors and matrices defined by

$$\tilde{c} = \begin{bmatrix} c \\ 0 \end{bmatrix} \in \mathbb{R}^{2s}, \; \tilde{Z} = \begin{bmatrix} Z \\ 0 \end{bmatrix} \in \mathbb{R}^{2s \times n}, \; \tilde{M} = \begin{bmatrix} M & 0 \\ 0 & |M| + 2|L| \end{bmatrix} \in \mathbb{R}^{2s \times 2s},$$

$$\tilde{L} = \begin{bmatrix} L & 0 \\ |L| & 0 \end{bmatrix} \in \mathbb{R}^{2s \times 2s}, \; \tilde{d} = \begin{bmatrix} d \\ 0 \end{bmatrix} \in \mathbb{R}^2, \; \tilde{a} = \begin{bmatrix} a & 0 \end{bmatrix} \in \mathbb{R}^{n \times 2}, \; \tilde{b} = \begin{bmatrix} b & 0 \\ 0 & |b| \end{bmatrix} \in \mathbb{R}^{2s \times 2}.$$

Then, Equations (19) and (20) yield

$$\begin{bmatrix} z \\ \delta z \end{bmatrix} = \begin{bmatrix} c \\ 0 \end{bmatrix} + \begin{bmatrix} Z \\ 0 \end{bmatrix} x + \begin{bmatrix} M & 0 \\ 0 & |M| + 2|L| \end{bmatrix} \begin{bmatrix} z \\ \delta z \end{bmatrix} + \begin{bmatrix} L & 0 \\ |L| & 0 \end{bmatrix} \begin{bmatrix} |z| \\ |\delta z| \end{bmatrix} = \begin{bmatrix} c + Zx + Mz + L|z| \\ (|M| + 2|L|)\delta z + |L||z| \end{bmatrix}$$

$$\begin{bmatrix} y \\ \delta y \end{bmatrix} = \tilde{d} + \tilde{a}^\top x + \tilde{b}^\top \tilde{z} = \begin{bmatrix} d \\ 0 \end{bmatrix} + \begin{bmatrix} a^\top x \\ 0 \end{bmatrix} + \begin{bmatrix} b & 0 \\ 0 & |b| \end{bmatrix}^\top \begin{bmatrix} z \\ \delta z \end{bmatrix} = \begin{bmatrix} d + a^\top x + b^\top z \\ |b|^\top \delta z \end{bmatrix} \,,$$

i.e., Equations (16) and (18). As can be seen, the matrices $\tilde{M}$ and $\tilde{L}$ have the required strictly lower triangular form. Furthermore, it is easy to check, that the switching depth of the abs-linear form of f carries over to the abs-linear form for $\check{f}$ in that also $(|\tilde{M}| + |\tilde{L}|)^\nu \neq 0 = (|\tilde{M}| + |\tilde{L}|)^{\nu+1}$. However, notice that this system is not reduced since the s radii are not switching variables, but globally nonnegative anyhow. We can now obtain explicit expressions for the central values, radii, and bounds for a given signature σ.

Corollary 1 (Explicit representation of the centered form). *For any definite signature $\sigma \neq 0$ and all $x \in \mathcal{P}_\sigma$ we have with $\Sigma = \mathrm{diag}(\sigma)$*

$$z_\sigma(x) = (I - M - L\Sigma)^{-1}(c + Zx) \quad \text{and} \quad |z_\sigma(x)| = \Sigma z_\sigma(x) \geq 0 \tag{21}$$

$$\delta z_\sigma(x) = (I - |M| - 2|L|)^{-1}|L|\Sigma(I - M - L\Sigma)^{-1}(c + Zx) \geq 0 \tag{22}$$

$$\nabla z_\sigma = (I - M - L\Sigma)^{-1} Z \implies \nabla_\sigma f = a^\top + b^\top (I - M - L\Sigma)^{-1} Z \tag{23}$$

$$\nabla \check{f}_\sigma = a^\top + \left[b^\top + |b|^\top (I - |M| - 2|L|)^{-1}|L|\Sigma \right] (I - M - L\Sigma)^{-1} Z \tag{24}$$

$$\nabla \hat{f}_\sigma = a^\top + \left[b^\top - |b|^\top (I - |M| - 2|L|)^{-1}|L|\Sigma \right] (I - M - L\Sigma)^{-1} Z, \tag{25}$$

where the restrictions of the functions and their gradients to $\mathcal{P}_\sigma$ are denoted by subscript σ. Notice that the gradients are constant on these open polyhedra.

Proof. Equations (21) and (23) follow directly from Equation (12), the abs-linear form (11) and the properties of Σ. Combining Equation (16) with (21) yields Equation (22). Since $\check{f}(x) = f(x) + \delta f(x)$ and $\hat{f}(x) = f(x) - \delta f(x)$, Equations (24) and (25) follow from the representation in abs-linear form and Equation (23). □

As one can see the computation of the gradient ∇f_σ requires the solution of one unit upper triangular linear system and that of both $\nabla \check{f}_\sigma$ and $\nabla \hat{f}_\sigma$ one more. Naturally, upper triangular systems are solved by back substitution, which corresponds to the reverse mode of algorithmic differentiation as described in the following section. Hence, the complexity for calculating the gradients is exactly the same as that for calculating the functions, which can be obtained by one forward substitution for f_σ and an extra one for δf_σ and thus $\check{f}_\sigma$ and $\hat{f}_\sigma$. The given $\nabla f_\sigma, \nabla \check{f}_\sigma$ and $\nabla \hat{f}_\sigma$ are proper gradients in the interior of the full dimensional domains $\mathcal{P}_\sigma$. For some or even many σ the inverse image $\mathcal{P}_\sigma$ of the map $x \mapsto \mathrm{sgn}(z(x))$ may be empty, in which case the formulas in the corollary do not apply. Checking the nonemptiness of $\mathcal{P}_\sigma$ for a given signature σ amounts to checking the consistency of a set of linear inequalities, which costs the same as solving an LOP and is thus nontrivial. Expressions for the generalized gradients at points in lower dimensional polyhedra are given in the following Section 4. There it is also not required that the abs-linear normal form has been reduced, but one may consider any given sequence of abs-linear operations.

The Two-Term Polyhedral Decomposition

It is well known ([15], Theorem 2.49) that all piecewise linear and globally convex or concave functions can be represented as the maximum or the minimum of a finite collection of affine functions, respectively. Hence, from the convex/concave decomposition we get the following drastic simplification of the classical min-max representation given, e.g., in [10].

Corollary 2 (Additive max/min decomposition of PL functions). *For every piecewise affine function $f : \mathbb{R}^n \mapsto \mathbb{R}$ there exist $k \geq 0$ affine functions $\alpha_i + a_i^\top x$ for $i = 1 \ldots k$ and $l \geq 0$ affine functions $\beta_j + b_j^\top x$ for $j = 1 \ldots l$ such that at all $x \in \mathbb{R}^n$*

$$f(x) = \underbrace{\max_{i=1 \ldots k}(\alpha_i + a_i^\top x)}_{\equiv \frac{1}{2}\check{f}(x)} + \underbrace{\min_{j=1 \ldots l}(\beta_j + b_j^\top x)}_{\equiv \frac{1}{2}\hat{f}(x)} \tag{26}$$

where furthermore $\hat{f}(x) \leq f(x) \leq \check{f}(x)$.

The max-part of this representation is what is called a polyhedral function in the literature [15]. Since the min-part is correspondingly the negative of a polyhedral function we may also refer to Equation (26) as a DP decomposition, i.e., the difference of two polyhedral functions.

We are not aware of a publication that gives a practical procedure for computing such a collection of affine functions $\alpha_i + a_i^\top x$, $i = 1 \ldots k$, and $\beta_j + b_j^\top x$, $j = 1 \ldots l$, for a given piecewise linear function f. Of course the critical question is in which form the function f is specified. Here as throughout our work we assume that it is given by a sequence of abs-linear operations. Then we can quite easily compute for each intermediate variable v representations of the form

$$v = \sum_{i=1}^{\tilde{m}} \max_{1 \leqslant j \leqslant k_i} (\alpha_{ij} + a_{ij}^\top x) + \sum_{i=1}^{\tilde{n}} \min_{1 \leqslant j \leqslant l_i} (\beta_{ij} + b_{ij}^\top x) \tag{27}$$

$$= \max_{\substack{j_i \in I_i \\ 1 \leqslant i \leqslant \tilde{m}}} \sum_{i=1}^{\tilde{m}} (\alpha_{ij_i} + a_{ij_i}^\top x) + \min_{\substack{j_i \in J_i \\ 1 \leqslant i \leqslant \tilde{n}}} \sum_{i=1}^{\tilde{n}} (\beta_{ij_i} + b_{ij_i}^\top x). \tag{28}$$

with index sets $I_i = \{1, \ldots, k_i\}$, $1 \leqslant i \leqslant \tilde{m}$, and $J_i = \{1, \ldots, l_i\}$, $1 \leqslant i \leqslant \tilde{n}$, since one has to consider all possibilities of selecting one affine function each from one of the $\tilde{m}$ max and $\tilde{n}$ min groups, respectively. Obviously, (28) involves $\prod_{i=1}^{m} k_i$ and $\prod_{i=1}^{n} \ell_i$ affine function terms in contrast to the first representation (27) which contains just $\sum_{i=1}^{m} k_i$ and $\sum_{i=1}^{n} \ell_i$ of them. Still the second version conforms to the classical representation of convex and concave piecewise linear functions, which yields the following result:

Corollary 3 (Explicit computation of the DP representation). *For any piecewise linear function given as abs-linear procedure one can explicitly compute the representation (26) by implementing the rules of Lemma 1.*

Proof. We will consider the representations (27) from which (26) can be directly obtained in the form (28). Firstly, the independent variables x_j are linear functions of themselves with gradient $a = e_j$ and inhomogeneity $\alpha = 0$. Then for multiplications by a constant $c > 0$ we have to scale all affine functions by c. Secondly, addition requires appending the expansions of the two summands to each other without any computation. Taking the negative requires switching the sign of all affine functions and interchanging the max and min group. Finally, to propagate through the absolute values we have to apply the rule (6), which means switching the signs in the min group, expressing it in terms of max and merging it with the existing max group. Here merging means pairwise joining each polyhedral term of the old max-group with each term in the switched min-group. Then the new min-group is the old one plus the old max-group with its sign switched. □

We see that taking the absolute value or, alternatively, maxima or minima generates the strongest growth in the number of polyhedral terms and their size. It seems clear that this representation is generally not very useful because the number of terms will likely blow up exponentially. This is not surprising because we will need one affine function for each element of the polyhedral decompositions of the domain of the max and min term. Typically, many of the affine terms will be redundant, i.e., could be removed without changing the values of the polyhedral terms. Unfortunately, identifying those already requires solving primal or dual linear programming problems, see, e.g., [16]. It seems highly doubtful that this would ever be worthwhile. Therefore, we will continue to advocate dealing with piecewise linear functions in a convenient procedural abs-linear representation.

4. Computation of Generalized Gradients and Constructive Oracle Paradigm

For optimization by variants of the DCA algorithm [17] one needs generalized gradients of the convex and the concave component. Normally, there are no strict rules for propagating generalized gradients through nonsmooth evaluation procedures. However, exactly this is simply assumed in the frequently invoked oracle paradigm, which states that at any point $x \in \mathbb{R}^n$ the function value

$f(x)$ and an element $g \in \partial f(x)$ can be evaluated. We have argued in [18] that this is not at all a reasonable assumption.

On the other hand, it is well understood that for the convex operations: Positive scaling, addition, and taking the maximum the rules are strict and simple. Moreover, then the generalized gradient in the sense of Clarke $\partial \check{f}(x) \subset \mathbb{R}^n$ is actually a subdifferential in that all its elements define supporting hyperplanes. Similarly $\partial \hat{f}(x)$ might be called a superdifferential in that the tangent planes bound the concave part from above.

In other words, we have at all $x \in \mathbb{R}^n$ and for all increments Δx

$$\check{f}(x + \Delta x) \geq \check{f}(x) + \check{g}^\top \Delta x \quad \text{if} \quad \check{g} \in \partial \check{f}(x)$$

and

$$\hat{f}(x + \Delta x) \leq \hat{f}(x) + \hat{g}^\top \Delta x \quad \text{if} \quad \hat{g} \in \partial \hat{f}(x),$$

which imply for $\check{g} \in \partial \check{f}(x)$ and $\hat{g} \in \partial \hat{f}(x)$ that

$$\hat{f}(x+\Delta x) + \check{f}(x) + \check{g}^\top \Delta x \leq 2f(x+\Delta x) \leq \check{f}(x+\Delta x) + \hat{f}(x) + \hat{g}^\top \Delta x, \tag{29}$$

where the lower bound on the left is a concave function and the upper bound is convex, both with respect to Δx. Notice that the generalized superdifferential $\partial \hat{f}$ being the negative of the subdifferential of $-\hat{f}$ is also a convex set.

Now the key question is how we can calculate a suitable pair of generalized gradients $(\check{g}, \hat{g}) \in \partial \check{f}(x) \times \partial \hat{f}(x)$. As we noted above the convex part and the negative of the concave part only undergo convex operations so that for $v = c\,u$

$$\partial \check{v} = \begin{cases} c\,\partial \check{u} & \text{if } c > 0 \\ 0 & \text{if } c = 0 \\ c\,\partial \hat{u} & \text{if } c < 0 \end{cases} \quad \text{and} \quad \partial \hat{v} = \begin{cases} c\,\partial \hat{u} & \text{if } c > 0 \\ 0 & \text{if } c = 0 \\ c\,\partial \check{u} & \text{if } c < 0 \end{cases} \tag{30}$$

and for $v = u + w$

$$\partial \check{v} = \partial \check{u} + \partial \check{w} \quad \text{and} \quad \partial \hat{v} = \partial \hat{u} + \partial \hat{w}. \tag{31}$$

Finally, for $v = |u|$ we find by Equation (6) that $\partial \hat{v} = \partial \hat{u} - \partial \check{u}$ as well as

$$\tfrac{1}{2}\partial \check{v} = \partial \max(\check{u}, -\hat{u}) = \begin{cases} \partial \check{u} & \text{if } u > 0 \\ \text{conv}\{\partial \check{u} \cup (-\partial \hat{u})\} & \text{if } u = 0 \\ -\partial \hat{u} & \text{if } u < 0 \end{cases}, \tag{32}$$

where we have used that $u = \tfrac{1}{2}(\check{u} + \hat{u})$ in Equation (32). The sign of the arguments u of the absolute value function are of great importance, because they determine the switching structure. For this reason, we formulated the cases in terms of u rather than in the convex/concave components. The operator conv$\{\cdot\}$ denotes taking the convex hull or envelope of a given usually closed set. It is important to state that within an abs-linear representation the multipliers c will stay constant independent of the argument x, even if they were originally computed as partial derivatives by an abs-linearization process and thus subject to round-off error. In particular their sign will remain fixed throughout whatever algorithmic calculation we perform involving the piecewise linear function f. So, actually the case $c=0$ could be eliminated by dropping this term completely and just initializing the left hand side v to zero.

Because we have set identities we can propagate generalized gradient pairs $(\nabla \check{u}, \nabla \hat{u}) \in \partial \check{u} \times \partial \hat{u}$ and perform the indicated algebraic operations on them, starting with the Cartesian basis vectors

$$\nabla \check{x}_j = \nabla \hat{x}_j = \nabla x_j = e_j \text{ since } \check{x}_j = \hat{x}_j = x_j \text{ for } j = 1 \ldots n.$$

The result of this propagation is guaranteed to be an element of $\partial \check{f} \times \partial \hat{f}$. Recall that in the merely Lipschitz continuous case generalized gradients cannot be propagated with certainty since for example the difference $v = w - u$ generates a proper inclusion $\partial v \subset \partial w - \partial u$. In that vein we must emphasize that the average $\frac{1}{2}(\nabla \check{f} + \nabla \hat{f})$ need not be a generalized gradient of $f = \frac{1}{2}(\check{f} + \hat{f})$ as demonstrated by the possibility that $\hat{f} = -\check{f}$ algebraically but we happen to calculate different generalized gradients of $\check{f}$ and $-\check{f}$ at a particular point x. In fact, if one could show that $\partial f = \frac{1}{2}(\partial \check{f} + \partial \hat{f})$ one would have verified the oracle paradigm, whose use we consider unjustified in practice. Instead, we can formulate another corollary for sufficiently piecewise smooth functions.

Definition 1. *For any $d \in \mathbb{N}$, the set of functions $f : \mathbb{R}^n \mapsto \mathbb{R}, y = f(x)$, defined by an abs-normal form*

$$\begin{aligned} z &= F(x, z, |z|), \\ y &= \varphi(x, z), \end{aligned}$$

with $F \in C^d(\mathbb{R}^{n+s+s})$ and $\varphi \in C^d(\mathbb{R}^{n+s})$, is denoted by $C^d_{abs}(\mathbb{R}^n)$.

Once more, this definition differs slightly from the one given in [7] in that y depends only on z and not on $|z|$ in order to match the abs-linear form used here. Then one can show the following result:

Corollary 4 (Constructive Oracle Paradigm). *For any function $f \in C^2_{abs}(\mathbb{R}^n)$ and a given point x there exist a convex polyhedral function $\widetilde{\Delta f}(x; \Delta x)$ and a concave polyhedral function $\widehat{\Delta f}(x; \Delta x)$ such that*

$$f(x + \Delta x) - f(x) = \frac{1}{2}\left(\widetilde{\Delta f}(x; \Delta x) + \widehat{\Delta f}(x; \Delta x)\right) + \mathcal{O}(\|\Delta x\|^2)$$

Moreover, both terms and their generalized gradients at $\Delta x = 0$ or anywhere else can be computed with the same order of complexity as f itself.

Proof. In [11], we show that

$$f(x + \Delta x) - f(x) = \Delta f(x; \Delta x) + \mathcal{O}(\|\Delta x\|^2),$$

where $\Delta f(x; \Delta x)$ is a piecewise linearization of f developed at x and evaluated at Δx. Applying the convex/concave decomposition of Theorem 1, one obtains immediately the assertion with a convex polyhedral function $\widetilde{\Delta f}(x; \Delta x)$ and a concave polyhedral function $\widehat{\Delta f}(x; \Delta x)$ evaluated at Δx. The complexity results follow from the propagation rules derived so far. □

We had hoped that it would be possible to use this approximate decomposition into polyhedral parts to construct at least locally an exact decomposition of a general function $f \in C^d_{abs}(\mathbb{R}^n)$ into a convex and compact part. The natural idea seems to add a sufficiently large quadratic term $\beta \|\Delta x\|^2$ to

$$f(x + \Delta x) - f(x) - \frac{1}{2}\widehat{\Delta f}(x; \Delta x) = \frac{1}{2}\widetilde{\Delta f}(x; \Delta x) + \mathcal{O}(\|\Delta x\|^2)$$

such that it would become convex. Then the same term could be subtracted from $\widehat{\Delta f}(x; \Delta x)$ maintaining its concavity. Unfortunately, the following simple example shows that this is not possible.

Example 1 (Half pipe). *The function*

$$f : \mathbb{R}^2 \mapsto \mathbb{R}, \qquad f(x_1, x_2) = \max(x_2^2 - \max(x_1, 0), 0) \qquad (33)$$

$$= \begin{cases} x_2^2 & \text{if } x_1 \leqslant 0 \\ x_2^2 - x_1 & \text{if } 0 \leqslant x_1 \leqslant x_2^2 \\ 0 & \text{if } 0 \leqslant x_2^2 \leqslant x_1 \end{cases},$$

in the class $\mathcal{C}_{abs}^\infty(\mathbb{R}^n)$ is certainly nonconvex as shown in Figure 1. As already observed in [19] this generally nonsmooth function is actually Fréchet differentiable at the origin $x = 0$ with a vanishing gradient $\nabla f(0) = 0$. Hence, we have $f(\Delta x) = \mathcal{O}(\|\Delta x\|^2)$ and may simply choose constantly $\widetilde{\Delta f}(0; \Delta x) \equiv 0 \equiv \widehat{\Delta f}(0; \Delta x)$. However, neither by adding $\beta \|\Delta x\|^2$ nor any other smooth function to $f(\Delta x)$ can we eliminate the downward facing kink along the vertical axis $\Delta x_1 = 0$. In fact, it is not clear whether this example has any DC decomposition at all.

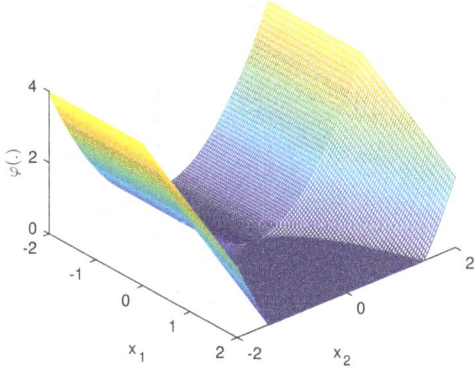

Figure 1. Half pipe example as defined in Equation (33).

Applying the Reverse Mode for Accumulating Generalized Gradients

Whenever gradients are propagated forward through a smooth evaluation procedure, i.e., for functions in $\mathcal{C}^2(\mathbb{R}^n)$, they are uniquely defined as affine combinations of each other, starting from Cartesian basis vectors for the components of x. Given only the coefficients of the affine combinations one can propagate corresponding adjoint values, or impact factors backwards, to obtain the gradient of a single dependent with respect to all independents at a small multiple of the operations needed to evaluate the dependent variable by itself. This cheap gradient result is a fundamental principle of computational mathematics, which is widely applied under various names, for example discrete adjoints, back propagation, and reverse mode differentiation. For a historical review see [20] and for a detailed description using similar notation to the current paper see our book [5]. For good reasons, there has been little attention to the reverse mode in the context of nonsmooth analysis, where one can at best obtain subgradients. The main obstacle is again that the forward propagation rules are only sharp when all elementary operations maintain convexity, which is by the way the only constructive way of verifying convexity for a given evaluation procedure. While general affine combinations and the absolute value are themselves convex functions, they do not maintain convexity when applied to a convex argument.

The last equation of Lemma 1 shows that one cannot directly propagate a subgradient of the convex radius functions δu because there is a reference to $v = |u|$ itself, which does not maintain convexity except when it is redundant due to its argument having a constant sign. However, it follows from the identity $\delta u = \frac{1}{2}(\check{u} - \hat{u})$ that for all intermediates u

$$\nabla \check{u} \in \partial \check{u} \wedge \nabla \hat{u} \in \partial \hat{u} \implies \tfrac{1}{2}(\nabla \check{u} - \nabla \hat{u}) \in \partial \delta u.$$

Hence one can get affine lower bounds of the radii, although one would probably prefer upper bounds to limit the discrepancy between the convex and concave parts. When $v = |u|$ and $u = 0$ we may choose according to Equation (32) any convex combination

$$\tfrac{1}{2} \nabla \check{v} = (1-\mu) \nabla \check{u} - \mu \nabla \hat{u} \quad \text{for} \quad 0 \le \mu \le 1. \tag{34}$$

It is tempting but not necessarily a good idea to always choose the weight μ equal to $\frac{1}{2}$ for simplicity.

Before discussing the reasons for this at the end of this subsection, let us note that from the values of the constants c, the intermediate values u, and the chosen weights μ it is clear how the next generalized gradient pair $(\nabla \check{v}, \nabla \hat{v})$ is computed as a linear combination of the generalized gradients of the inputs for each operation, possibly with a switch in their roles. That means after only evaluating the function f itself, not even the bounds $\check{f}$ and $\hat{f}$, we can compute a pair of generalized gradients in $\partial \check{f} \times \partial \hat{f}$ using the reverse mode of algorithmic differentiation, which goes back to at least [21] though not under that name. The complexity of this computation will be independent of the number of variables and relative to the complexity of the function f itself. All the operations are relatively benign, namely scaling by constants, interchanges and additions and subtractions. After all the reverse mode is just a reorganization of the linear algebra in the forward propagation of gradients. Hence, it appears that we can be comparatively optimistic regarding the numerical stability of this process.

To be specific we will indicate the (scalar) adjoint value of all intermediates $\check{u}$ and $\hat{u}$ as usual by $\bar{\check{u}} \in \mathbb{R}$ and $\bar{\hat{u}} \in \mathbb{R}$. They are all initialized to zero except for either $\bar{\check{y}} = 1$ or $\bar{\hat{y}} = 1$. Then at the end of the reverse sweep, the vectors $(\bar{x}_j)_{j=1}^n$ represent either $\nabla \check{y}$ or $\nabla \hat{y}$, respectively. For computational efficiency one may propagate both adjoint components simultaneously, so that one computes with sextuplets consisting of $\check{u}, \hat{u}$ and their adjoints with respect to $\check{y}$ and $\hat{y}$. In any case we have the following adjoint operations. For $v = u + w$

$$(\bar{\check{w}}, \bar{\hat{w}}) \mathrel{+}= (\bar{\check{v}}, \bar{\hat{v}}) \quad \text{and} \quad (\bar{\check{u}}, \bar{\hat{u}}) \mathrel{+}= (\bar{\check{v}}, \bar{\hat{v}}),$$

for $v = cu$

$$(\bar{\check{u}}, \bar{\hat{u}}) \mathrel{+}= \begin{cases} c(\bar{\check{v}}, \bar{\hat{v}}) & \text{if } c > 0 \\ (0,0) & \text{if } c = 0, \\ c(\bar{\hat{v}}, \bar{\check{v}}) & \text{if } c < 0 \end{cases}$$

and finally for $v = |u|$

$$(\bar{\check{u}}, \bar{\hat{u}}) \mathrel{+}= \begin{cases} (2\bar{\check{v}} - \bar{\hat{v}}, \bar{\hat{v}}) & \text{if } u > 0 \\ (-\bar{\hat{v}} + 2(1-\mu)\bar{\check{v}}, \bar{\hat{v}} - 2\mu\bar{\check{v}}) & \text{if } u = 0. \\ (-\bar{\hat{v}}, \bar{\check{v}} - 2\bar{\check{v}}) & \text{if } u < 0 \end{cases} \tag{35}$$

Of course, the update for the critical case $u = 0$ of the absolute value is just the convex combination for the two cases $u > 0$ and $u < 0$ weighted by μ. Due to round-off errors it is very unlikely that the critical case $u = 0$ ever occurs in floating point arithmetic. Once more, the sign of the arguments u of the absolute value function are of great importance, because they determine on which faces of the polyhedral functions $\check{f}$ and $\hat{f}$ the current argument x is located. In some situations one prefers a gradient that is limiting in that it actually occurs as a proper gradient on one of the adjacent smooth pieces. For example, if we had simply $f(x) = v = |x|$ for $x \in \mathbb{R}$ and chose $\mu = \frac{1}{2}$ we would get

$\check{v} = 2|x|, \hat{v} = 0$ and find by Equation (34) that $\nabla \check{v} = 2(\frac{1}{2} - \frac{1}{2}) = 0$ at $x = \check{x} = \hat{x} = 0$. This is not a limiting gradient of $\check{v}$ since $\partial \check{v} = [-2, 2]$, whose interior contains the particular generalized gradient 0.

5. Exploiting the Convex/concave Decomposion for the DC Algorithm

In order to minimize the decomposed objective function f we may use the DCA algorithm [17] which is given in its basic form using our notation by

Choose $x_0 \in \mathbb{R}^n$
For $k = 0, 1, 2, \ldots$
  Calculate $g_k \in -\partial(\frac{1}{2}\hat{f})(x_k)$
  Calculate $x_{k+1} \in \partial(\frac{1}{2}\check{f})^*(g_k)$

where $(\frac{1}{2}\check{f})^*$ denotes the Fenchel conjugate of $(\frac{1}{2}\check{f})$. For a convex function $h: \mathbb{R}^n \mapsto \mathbb{R}$ one has

$$w \in \partial h^*(y) \quad \Leftrightarrow \quad w \in \underset{x \in \mathbb{R}^n}{\operatorname{argmin}}\{h(x) - y^\top x\},$$

see [15], Chapter 11. Hence, the classic DCA reduces in our Euclidean scenario to a simple recurrence

$$x_{k+1} \in \underset{x \in \mathbb{R}^n}{\operatorname{argmin}}\left\{\check{f}(x) + \hat{g}_k^\top x\right\} \quad \text{for some} \quad \hat{g}_k \in \partial \hat{f}(x_k). \tag{36}$$

The objective function on the left hand side is a constantly shifted convex polyhedral upper bound on $2f(x)$ since

$$\check{f}(x) + \hat{g}_k^\top x = 2f(x) - \left(\hat{f}(x) - \hat{g}_k^\top x\right) \geq 2f(x) - \hat{f}(x_k) + \hat{g}_k^\top x_k. \tag{37}$$

It follows from Equation (29) and x_{k+1} being a minimizer that

$$f(x_{k+1}) \leq \tfrac{1}{2}\left(\check{f}(x_{k+1}) + \hat{f}(x_k) + \hat{g}_k^\top(x_{k+1} - x_k)\right)$$
$$\leq \tfrac{1}{2}\left(\check{f}(x_k) + \hat{f}(x_k)\right) = f(x_k).$$

Now, since (36) is an LOP, an exact solution x_{k+1} can be found in finitely many steps, for example by a variant of the Simplex method. Moreover, we can then assume that x_{k+1} is one of finitely many vertex points of the epigraph of $\check{f}$. At these vertex points, f itself attains a finite number of bounded values. Provided f itself is bounded below, we can conclude that for any choice of the $\hat{g}_k \in \partial \hat{f}_{\sigma(k)}$ the resulting function values $f(x_k)$ can only be reduced finitely often so that $f(x_k) = f(x_{k-1})$ and w.l.o.g. $x_k = x_{k-1}$ eventually. We then choose the next $\hat{g}_k = \nabla \hat{f}_{\sigma^{(k)}}$ with $\sigma^{(k)} = \sigma^{(k-1)} \triangleright \sigma(x_k)$ as the reflection of $\sigma^{(k-1)}$ at $\sigma(x_k)$ as defined in (15). If then again $f(x_{k+1}) = f(x_k)$ it follows from Corollary A2 that x_k is a local minimizer of f and we may terminate the optimization run. Hence we obtain the DCA variant listed in Algorithm 1, which is guaranteed to reach local optimality under LIKQ. It is well defined even without this property and we conjecture that otherwise the final iterate is still a stationary point of f. The path of the algorithm on the example discussed in Section 5 is sketched in Figure 3. It reaches the stationary point $(0, -1)$ where $\sigma = (0, -1, 0)$ from within the polyhedron with the signature $(-1, -1, -1)$ and then continues after the reflection $(1, -1, 1) = (-1, -1, -1) \triangleright (0, -1, 0)$. From within that polyhedron the inner loop reaches the point $(1, 1)$ with signature $(1, 0, 0)$, whose minimality is established after a search in the polyhedron $\overline{\mathcal{P}}_{(1,1,-1)}$.

If the function $f(x)$ is unbounded below, so will be one of the inner convex problems and the convex minimizer should produce a ray of infinite descent instead of the next iterate x_{k+1}. This exceptional scenario will not be explicitly considered in the remainder of the paper. The reflection operation is designed to facilitate further descent or establish local optimality. It is discussed in the context of general optimality conditions in the following subsection.

Algorithm 1 Reflection DCA

Require: $x_0 \in \mathbb{R}^n$,
1: Set $f_{-1} = \infty$ and Evaluate $f_0 = f(x_0)$
2: **for** $k = 0, 1, \ldots$ **do**
3:     **if** $f_k < f_{k-1}$ **then**   ▷ Normal iteration with function reduction
4:         Choose $0 \notin \sigma > \sigma(x_k)$ ▷ Here different heuristics may be applied
5:         Compute $\hat{g}_k = \nabla \hat{f}_\sigma$ ▷ Apply formula of Corollary 1
6:     **else**   ▷ The starting point was already optimal
7:         Reflect $\check{\sigma} = \sigma \triangleright \sigma(x_k)$ ▷ The symbol $\triangleright$ is defined in Equation (15).
8:         Update $\hat{g}_k = \nabla \hat{f}_{\check{\sigma}}$
9:     **end if**
10:     Calculate $x_{k+1} \in \text{argmin} \left\{ \check{f}(x) + \hat{g}_k^\top x \,\middle|\, x \in \mathbb{R}^n \right\}$ ▷ Apply any LOP finite solver
11:     Set $f_{k+1} = f(x_{k+1})$
12:     **if** $f_{k+1} = f_k = f_{k-1}$ **then**   ▷ Local optimality established
13:         Stop
14:     **end if**
15: **end for**

5.1. Checking Optimality Conditions

Stationarity of x_k happens when the convex function $\check{f}(x) + \hat{g}_k^\top x$ is minimal at x_k so that for all large k

$$0 \in \partial \check{f}(x_k) + \hat{g}_k \iff \hat{g}_k \in \partial \check{f}(x_k) \cap (-\partial \check{f}(x_k)) \neq \emptyset. \tag{38}$$

The nonemptiness condition on the right hand side is known as criticality of the DC decomposition at x_k, which is necessary but not sufficient even for local optimality of $f(x)$ at x_k. To ensure the latter one has to verify that all $\hat{g}_k \in \partial \hat{f}(x_k)$ satisfy the criticality condition (38) so that

$$\partial \hat{f}(x_k) \subset -\partial \check{f}(x_k) \iff \partial^L \hat{f}(x_k) \subset -\partial \check{f}(x_k). \tag{39}$$

The left inclusion is a well known local minimality condition [22], which is already sufficient in the piecewise linear case. The right inclusion is equivalent to the left one due to the convexity of $\partial \check{f}(x_k)$.

If $\check{f}$ and $\hat{f}$ were unrelated convex and concave polyhedral functions, one would normally consider it extremely unlikely that $\hat{f}$ were nonsmooth at any one of the finitely many vertices of the polyhedral domain decomposition of $\check{f}$. For instance when $\hat{f}$ is smooth at x_k we find that $\partial \hat{f}(x_k) = \{\hat{g}_k\}$ is a singleton so that criticality according to Equation (38) is already sufficient for local minimality according to Equation (39). As we have seen in Theorem 1 the two parts have exactly the same switching structure. That means they are nonsmooth on the same skeleton of lower dimensional polyhedra. Hence, neither $\partial^L \check{f}(x_k)$ nor $\partial^L \hat{f}(x_k)$ will be singletons at minimizing vertices of the upper bound so that checking the validity of Equation (39) appears to be a combinatorial task at first sight.

However, provided the Linear Independence Kink Qualification (LIKQ) defined in [7] is satisfied at the candidate minimizer x_k, the minimality can be tested with cubic complexity even in case of a dense abs-linear form. Moreover, if the test fails one can easily calculate a descent direction d. The details of the optimality test in our context including the calculation of a descent direction are given in the Appendix A. They differ slightly from the ones in [7]. Rather than applying the optimality test Proposition A1 explicitly, one can use its Corollary A2 stating that if $\hat{x}$ with $\hat{\sigma} = \sigma(\hat{x})$ is a local minimizer of the restriction of f to a polyhedron $\mathcal{P}_\sigma$ with definite $\sigma > \hat{\sigma}$ then it is a local minimizer of the unrestricted f if and only if it also minimizes the restriction of f to $\bar{\mathcal{P}}_{\check{\sigma}}$ with the reflection $\check{\sigma} = \sigma \triangleright \hat{\sigma}$. The latter condition must be true if $\hat{x}$ also minimizes $f(x) + \nabla \hat{f}_{\check{\sigma}}$, which can be checked by solving that convex problem. If that test fails the optimization can continue.

5.2. Proximal Rather Than Global

By some authors the DCA algorithm has been credited with being able to reach global minimizers with a higher probability than other algorithms. There is really no justification for this optimism in the light of the following observation. Suppose the objective $f(x) = \frac{1}{2}(\check{f}(x) + \hat{f}(x))$ has an isolated local minimizer x_*. Then there exists an $\varepsilon > 0$ such that the level set $\{x \in \mathbb{R}^n : f(x) \leq f(x_*) + \varepsilon\}$ has a bounded connected component containing x_*, say $\mathcal{L}_\varepsilon$. Now suppose DCA is started from any point $x_0 \in \mathcal{L}_\varepsilon$. Since $f_0(x) \equiv \frac{1}{2}(\check{f}(x) + \hat{f}(x_0) + \hat{g}(x_0)^\top (x - x_0))$ is by Equation (37) a convex upper bound on $f(x)$ its level set $\{f_0(x) \leq f(x_0)\}$ will be contained in $\mathcal{L}_\varepsilon$. Hence any step from x_0 that reduces the upper bound $f_0(x)$ must stay in the same component, so there is absolutely no chance to move away from the catchment $\mathcal{L}_\varepsilon$ of x_0 towards another local minimizer of f, whether global or not. In fact, by adding the convex term

$$\tfrac{1}{2}\left(\hat{f}(x_0) + \hat{g}(x_0)^\top (x - x_0) - \hat{f}(x)\right) \geq 0,$$

which vanishes at x_0, to the actual objective $f(x)$ one performs a kind of regularization, like in the proximal point method. This means the step is actually held back compared to a larger step that might be taken by a method that only requires the reduction of $f(x)$ itself.

Hence we may interpret DCA as a proximal point method where the proximal term is defined as an affinely shifted negative of the concave part. Since in general the norm and the coefficient defining the proximal term may be quite hard to select, this way of defining it may make a lot of sense. However, it is certainly not global optimization. Notice that in this argument we have used neither the polyhedrality nor the inclusion property. So it applies to a general DC decomposition on Euclidean space. Another conclusion from the "holding back" observation is that it is probably not worthwhile to minimize the upper bound very carefully. One might rather readjust the shift $\hat{g}^\top x$ after a few or even just one iteration.

6. Nesterov's Piecewise Linear Example

According to [6], Nesterov suggested three Rosenbrock-like test functions for nonsmooth optimization. One of them given by

$$f(x) = \tfrac{1}{4}|x_1 - 1| + \sum_{i=1}^{n-1} |x_{i+1} - 2|x_i| + 1| \tag{40}$$

is nonconvex and piecewise linear. It is shown in [6] that this function has 2^{n-1} Clarke stationary points only one of which is a local and thus the global minimizer. Numerical studies showed that optimization algorithms tend to be trapped at one of the stationary points making it an interesting test problem. We have demonstrated in [23] that using an active signature strategy one can guarantee convergence to the unique minimizer from any starting point albeit using in the worst case 2^n iterations as all stationary points are visited. Let us first write the problem in the new abs-linear form.

Defining the $s = 2n$ switching variables

$$z_i = F_i(x, |z|) = x_i \quad \text{for} \quad 1 \leq i < n, \qquad z_n = F_n(x, |z|) = x_1 - 1,$$

and

$$z_{n+i} = F_{n+i}(x, |z|) = x_{i+1} - 2|z_i| + 1 \text{ for } 1 \leq i < n, \quad z_s = \tfrac{1}{4}|z_n| + \sum_{i=1}^{n-1} |z_{n+i}|$$

the resulting objective function is then simply identical to $y = f(x) = z_s$. With the vectors and matrices

$$c^\top = (0, -1, e_{n-1}^\top, 0) \in \mathbb{R}^{(n-1)+1+(n-1)+1}, \quad Z = \begin{bmatrix} I_{n-1} & 0 \\ I_{n-1} & 0 \\ 0 & 1 \\ 0 & 0 \end{bmatrix} \in \mathbb{R}^{s \times (n-1)+1},$$

$$M = 0, \quad L = \begin{bmatrix} 0 & 0 & 0 & 0 \\ 0 & 0 & 0 & 0 \\ -2\,I_{n-1} & 0 & 0 & 0 \\ 0 & \frac{1}{4} e_{n-1}^\top & 0 \end{bmatrix} \in \mathbb{R}^{s \times (n-1)+1+(n-1)+1}, \quad d = 0 \in \mathbb{R},$$

$$a = 0, \quad b^\top = (0, \cdots, 0, 1) \in \mathbb{R}^{(2n-1)+1},$$

where Z and L have different row partitions, one obtains an abs-linear form (11) of f. Here, I_k denotes the identity matrix of dimension k, $e^\top = (1, \cdots, 1) \in \mathbb{R}^k$ the vector containing only ones and the symbol 0 pads with zeros to achieve the specified dimensions. One can easily check that $|L|^2 \neq 0 = |L|^3$, hence this example has switching depth $\nu = 2$. The geometry of the situation is depicted in Figure 3, which was already briefly discussed in Sections 3 and 5.

Since the corresponding extended abs-linear form for $\tilde{f} = (y, \delta y)$ does not provide any new insight we do not state it here. Directly in terms of the original equations we obtain for the radii

$$\delta z_i = 0 \text{ for } 1 \leq i \leq n, \quad \delta z_{n+i} = 2|z_i| = 2|x_i| \text{ for } 1 \leq i < n \tag{41}$$

and

$$\begin{aligned} \delta f = \delta z_s &= \tfrac{1}{4}|z_n| + \sum_{i=1}^{n-1}(|z_{n+i}| + 2\delta z_{n+i}) \\ &= \tfrac{1}{4}|x_1 - 1| + \sum_{i=1}^{n-1}(|x_{i+1} - 2|x_i| + 1| + 4|x_i|). \end{aligned} \tag{42}$$

Thus, from Equation (7) we get the convex and concave part explicitly as

$$\left. \begin{aligned} \check{z}_i &= z_i = \hat{z}_i \text{ for } 1 \leq i \leq n, \\ \check{z}_{n+i} &= x_{i+1} + 1 \\ \hat{z}_{n+i} &= x_{i+1} - 4|z_i| + 1 = x_{i+1} - 4|x_i| + 1 \end{aligned} \right\} \text{ for } 1 \leq i < n$$

and most importantly

$$\check{f} = z_s + \delta z_s = \tfrac{1}{2}|x_1 - 1| + 2 \sum_{i=1}^{n-1} (|x_{i+1} - 2|x_i| + 1| + 2|x_i|)$$

$$\hat{f} = z_s - \delta z_s = -4 \sum_{i=1}^{n-1} |x_i|.$$

Clearly $\hat{f}$ is a concave function and to check the convexity of $\check{f}$ we note that

$$\begin{aligned} |x_{i+1} - 2|x_i| + 1| + 2|x_i| &= |2|x_i| - 1 - x_{i+1}| + (2|x_i| - 1 - x_{i+1}) + x_{i+1} + 1 \\ &= 1 + x_{i+1} + 2\max(0, 2|x_i| - x_{i+1} - 1). \end{aligned} \tag{43}$$

The last expression is the sum of an affine function and the positive part of the sum of the absolute value and an affine function, which must therefore also be convex. The corresponding term in

Equation (42) is the same with the convex function $2|x_i|$ added, so that δf is also convex in agreement with the general theory. Finally, one verifies easily that

$$\hat{f} \leq f = \tfrac{1}{2}(\check{f} + \hat{f}) \leq \check{f},$$

which is the whole idea of the decomposition. It would seem that the automatic decomposition by propagation through the abs-linear procedure yields a rather tight result. The function f as well as the lower and upper bound given by the convex/concave decomposition are illustrated on the left hand side of Figure 2. Notice that the switching structure is indeed identical for all three as stated in Theorem 1. On the right hand side of Figure 2, the difference $2\delta f$ between the upper, convex and lower, concave bound is shown, which is indeed convex.

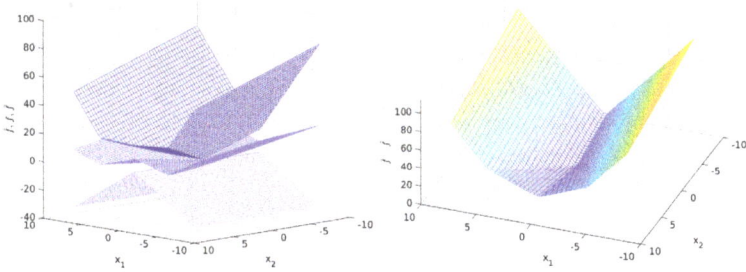

Figure 2. Nesterov–Rosenbrock test function polyhedral inclusion for $n = 2$.

It is worthwhile to look at the condition number of the decomposition, namely we get the following trivial bound

$$\begin{aligned}
\kappa(\check{f}, \hat{f}) &= \sup_{x \in \mathbb{R}^n} \frac{\tfrac{1}{2}|x_1 - 1| + 2\sum_{i=1}^{n-1}\left(\left|x_{i+1} - 2|x_i| + 1\right| + 4|x_i|\right)}{\tfrac{1}{2}|x_1 - 1| + 2\sum_{i=1}^{n-1}|x_{i+1} - 2|x_i| + 1|} \\
&= 1 + \sup_{x \in \mathbb{R}^n} \frac{8\sum_{i=1}^{n-1}|x_i|}{\tfrac{1}{4}|x_1 - 1| + 2\sum_{i=1}^{n-1}|x_{i+1} - 2|x_i| + 1|} = \infty.
\end{aligned}$$

The disappointing right hand side value follows from the fact that at the well known unique global optimizer $x_* = (1, 1, \ldots, 1) \in \mathbb{R}^n$ the numerator is zero and the denominator positive. However, elsewhere, we can bound the conditioning as follows.

Lemma 3. *In case of the example* (40) *there is a constant* $c \in \mathbb{R}$ *such that*

$$\kappa(\check{f}(x), \hat{f}(x)) \leq 1 + \frac{c}{\min(\|x - x_*\|, 3)} \,. \tag{44}$$

Proof. Since the denominator is piecewise linear and vanishes only at the minimizer x_* there must be a constant $c_0 > 0$ such that for $\|x - x_*\|_\infty \leq 3$

$$\frac{8\sum_{i=1}^{n-1}|x_i|}{\tfrac{1}{4}|x_1 - 1| + 2\sum_{i=1}^{n-1}|x_{i+1} - 2|x_i| + 1|} \leq \frac{8\sum_{i=1}^{n-1}|x_i|}{c_0 \|x - x_*\|_\infty} \leq \frac{8(n-1)\|x\|_\infty}{c_0\|x - x_*\|_\infty} \leq \frac{32(n-1)}{c_0\|x - x_*\|_\infty},$$

which takes the value $32(n-1)/(3c_0)$ on the boundary. On the other hand we get for $\|x\|_\infty \geq 2$ and thus in particular $\|x - x_*\|_\infty \geq 3$

$$\frac{8\sum_{i=1}^{n-1}|x_i|}{\tfrac{1}{4}|x_1 - 1| + 2\sum_{i=1}^{n-1}|x_{i+1} - 2|x_i| + 1|} \leq \frac{4(n-1)\|x\|_\infty}{\max_{1 \leq i < n} |2|x_i| - x_{i+1} - 1|} \leq \frac{4(n-1)}{2 - 1 - 1/2} \leq 8(n-1).$$

Assuming without loss of generality that $c_0 \leqslant 4/3$ we can combine the two bounds to obtain the assertion with $c \equiv 32(n-1)/c_0$. □

Hence, we see the condition number $\kappa(\check{f}(x), \hat{f}(x))$ is nicely bounded and the decomposition should work as long as our optimization algorithm has not yet reached its goal x_*. It is verified in the companion article [24], that the DCA exploiting the observations made in this paper reaches the global minimizer in finitely many steps. It was already shown in [7] that the LIKQ condition is satisfied everywhere and that the optimality test singles out the unique minimizer correctly. In Figure 3, the arrows indicate the path of our reflection version of the DCA method as described in Section 5.

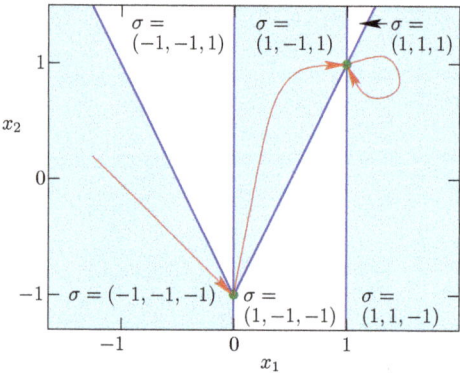

Figure 3. Signatures and reflection-based DCA for Nesterov–Rosenbrock variant (40) with $n = 2$.

7. Summary, Conclusions and Outlook

In this paper the following new results were achieved

- For every piecewise linear function f given as an abs-linear evaluation procedure, rules for simultaneously evaluating its representation as the average of a concave lower bound $\hat{f}$ and a convex upper bound $\check{f}$ are derived.
- The two bounds can be constructively expressed as a single maximum and minimum of affine functions, which drastically simplifies the classical min – max representation. Due to its likely combinatorial complexity we do not recommend this form for practical calculations.
- For the two bounds $\check{f}$ and $\hat{f}$, generalized gradients $\check{g}$ and $\hat{g}$ can be propagated forward or reverse through the convex or concave operations that define them. The gradients are not unique but guaranteed to yield supporting hyperplanes and thus provide a verified version of the oracle paradigm.
- The DCA algorithm can be implemented such that a local minimizer is reached in finitely many iterations, provided the Linear Independence Kink Qualification (LIKQ) is satisfied. It is conjectured that without this assumption the algorithm still converges in finitely many steps to a Clarke stationary point. Details on this can be found in the companion paper [24].

These results are illustrated on the piecewise linear Rosenbrock variant of Nesterov.

On a theoretical level it would be gratifying and possibly provide additional insight, to prove the result of Corollary A3 directly using the explicit representations of the generalized differentials of the convex and concave part given in Corollary 1. Moreover, it remains to be explored what happens when LIKQ is not satisfied. We have conjectured in [25] that just verifying the weaker Mangasarian Fromovitz Kink Qualification (MFKQ) represents an NP hard task. Possibly, there are other weaker conditions that can be cheaply verified and facilitate the testing for at least local optimality.

Global optimality can be characterized theoretically in terms of ε-subgradients, albeit with ε arbitrarily large [26]. There is the possibility that the alternative definition of ε-gradients given in [18] might allow one to constructively check for global optimality. It does not really seem clear how these global optimality conditions can be used to derive corresponding algorithms.

The implementation of the DCA algorithm can be optimized in various ways. Notice that for applying the Simplex method in standard form, one could use for the representation as DC function the max-part in the more economical representation Equation (27) introducing $\bar{m}$ additional variables, rather than the potentially combinatorial Equation (28) to assemble the constraint matrix. In any case it seems doubtful that solving each sub problem to completion is a good idea, especially as the resulting step in the outer iteration is probably much too small anyhow. Therefore, the generalized gradient of the concave part, which defines the inner problem, should probably be updated much more frequently. Moreover, the inner solver might be an SQOP type active signature method or a matrix free gradient method with momentum term, as is used in machine learning, notwithstanding the nonsmoothness of the objective. Various options in that range will be discussed and tested in the companion article [24].

Finally, one should always keep in mind that the task of minimizing a piecewise linear function will most likely occur as an inner problem in the optimization of a piecewise smooth and nonlinear function. As we have shown in [27] the local piecewise linear model problem can be obtained easily by a slight generalization of automatic or algorithmic differentiation, e.g., ADOL-C [28] and Tapenade [29].

Author Contributions: Conceptualization, A.G. and A.W.; methodology, A.G. and A.W.; writing–original draft preparation; writing–review and editing, A.G. and A.W. Both authors have read and agreed to the published version of the manuscript.

Funding: We acknowledge support by the German Research Foundation (DFG) and the Open Access Publication Fund of Humboldt-Universität zu Berlin.

Acknowledgments: We thank Napsu Karmitsa and Sona Taheri for inviting us to participate in this special issue in honor of Adil M. Bagirov. We also thank the three anonymous referees, who asked for various corrections and clarifications, which made the paper much more self-contained and readable.

Conflicts of Interest: The authors declare no conflict of interest.

Appendix A. Polynomial Optimality Test Based on Abs-Linear Form

As illustrated for the Nesterov test function, it may be advantageous to use intermediate variables z_i that are not arguments of the absolute value themselves. For simplicity, we assume that these switching variables that do not impose nonsmoothness are located in the last components of z and that only the $\tilde{s} \leqslant s$ components $z_1, \ldots z_{\tilde{s}}$ are arguments of the absolute value. Let us abbreviate the current iterate x_k with $\mathring{x} \equiv x_k$ and denote the corresponding switching vector by $\mathring{z} = z(\mathring{x})$, the signature vector $\mathring{\sigma} = \text{sgn}(\mathring{z})$ and the active index set by $\alpha \equiv \{i \leqslant \tilde{s} : \mathring{\sigma}_i = 0\}$ with cardinality $m \equiv |\alpha| \leqslant \tilde{s}$. Consequently, there are exactly 2^m definite signatures by $\sigma > \mathring{\sigma}$ and the same number of limiting gradients for the three generalized differentials $\partial \breve{f}, \partial \hat{f}$, and ∂f.

For all $x \in \mathcal{P}_{\mathring{\sigma}}$, the signature $\mathring{\sigma}$ is constant and we can use Corollary 1 to define the smooth function

$$z_{\mathring{\sigma}}(x) = (I - M - L\mathring{\Sigma})^{-1}(c + Zx) = \mathring{c} + \mathring{Z}x, \tag{A1}$$

where we have pulled out the unit lower triangular factor $(I - M - L\mathring{\Sigma})$ such that

$$\mathring{Z} = (I - M - L\mathring{\Sigma})^{-1}Z \quad \text{and} \quad \mathring{c} = (I - M - L\mathring{\Sigma})^{-1}c.$$

For $x \approx \mathring{x}$ to be contained in the extended closure $\bar{\mathcal{P}}_{\mathring{\sigma}}$ as defined in Equation (14), it must satisfy the m linear equations

$$P_\alpha z(x) = 0 \in \mathbb{R}^m \quad \text{for} \quad P_\alpha = (e_i^\top)_{i \in \alpha} \in \mathbb{R}^{m \times \tilde{s}}$$

with e_i denoting the ith unit vector in $\mathbb{R}^{\tilde{s}}$. Thus it is necessary and sufficient for $\bar{\mathcal{P}}_{\mathring{\sigma}}$ to be a polyhedron of dimension $n - m$ that the Jacobian $P_\alpha \mathring{Z} \in \mathbb{R}^{m \times n}$ has full row rank m. This rank condition was introduced as LIKQ in [7] and obviously requires that no more than n switches are active at $\mathring{x}$. As discussed in [7], for the point $\mathring{x}$ to be a local minimizer of f it is necessary that it solves the trunk problem

$$\min a^\top x + b^\top z \quad \text{s.t.} \quad |\mathring{\Sigma}|z - \mathring{c} - \mathring{Z}x = 0 \ .$$

Here $|\mathring{\Sigma}| \in \mathbb{R}^{\mathring{s} \times \mathring{s}}$ is the projection onto the $\mathring{s} - m$ vector components whose indices do not belong to α so the equality constraint combines (A1) and the constraint $P_\alpha z = 0$. Now we get from KKT theory or equivalently LOP duality that $\mathring{x}$ is a minimizer on $\mathcal{P}_\alpha$ if and only if for some Lagrange multiplier vector $\lambda \in \mathbb{R}^{\mathring{s}}$

$$a^\top = -\lambda^\top \mathring{Z} \quad \text{and} \quad b^\top = \lambda^\top |\mathring{\Sigma}| \ . \tag{A2}$$

Since $I = |\mathring{\Sigma}| + P_\alpha^\top P_\alpha$ we derive that

$$\lambda^\top (I - |\mathring{\Sigma}|)\mathring{Z} = \lambda_\alpha^\top P_\alpha \mathring{Z} = -a^\top - b^\top \mathring{Z} \ . \tag{A3}$$

where $\lambda_\alpha \equiv P_\alpha \lambda$. This is a generally overdetermined system of n equations in the m components of λ_α. If it is solvable the full multiplier vector $\lambda = P_\alpha^\top \lambda_\alpha + |\mathring{\Sigma}| b$ is immediately available. Because of the assumed full rank of the Jacobian $P_\alpha \mathring{Z}$ we have $m \leq n$, and if $\mathring{x}$ is a vertex in that $m = n$ the tangential stationarity condition (A3) is automatically satisfied.

Now it is necessary and sufficient for local minimality that $\mathring{x}$ is also a minimizer of f on all polyhedra $\bar{\mathcal{P}}_\sigma$ with definite $\sigma > \mathring{\sigma}$. Any such $\sigma > \mathring{\sigma}$ can be written as $\sigma = \mathring{\sigma} + \gamma$ with $\gamma \in \{-1, 0, 1\}^{\mathring{s}}$ structurally orthogonal to $\mathring{\sigma}$ such that for $\Gamma = \text{diag}(\gamma)$ we have the matrix equations

$$\Sigma = \mathring{\Sigma} + \Gamma \quad \text{and} \quad \mathring{\Sigma}\Gamma = 0 = |\mathring{\Sigma}|\Gamma \ .$$

Then we can express the $z(x) = z_\sigma(x)$ for $x \in \mathcal{P}_\sigma$ as

$$z_\sigma(x) = z_{\mathring{\sigma}+\gamma}(x) = (I - M - L\mathring{\Sigma} - L\Gamma)^{-1}(c + Zx)$$
$$= (I - \mathring{L}\Gamma)^{-1}(\mathring{c} + \mathring{Z}x) \ ,$$

with $\mathring{L} \equiv (I - M - L\mathring{\Sigma})^{-1}L$. Now $\mathring{x}$ must be the minimizer of f on $\bar{\mathcal{P}}_\sigma$, i.e., solve the problem

$$\min a^\top x + b^\top z \quad \text{s.t.} \quad (I - \mathring{L}\Gamma)z = \mathring{c} + \mathring{Z}x, \quad P_\alpha \Gamma z \geq 0 \in \mathbb{R}^m \ . \tag{A4}$$

Notice that the inequalities are only imposed on the sign constraints that are active at $\mathring{x}$ since the strict inequalities are maintained in a neighborhood of $\mathring{x}$ due to the continuity of $z(x)$. Then we get again from KKT theory or equivalently LOP duality that still $a^\top = -\lambda^\top \mathring{Z}$ and for a second multiplier vector $0 \leq \mu \in \mathbb{R}^m$ the equalities

$$a^\top = -\lambda^\top \mathring{Z} \quad \text{and} \quad b^\top = \lambda^\top (I - \mathring{L}\Gamma) + \mu^\top P_\alpha \Gamma \ . \tag{A5}$$

Multiplying from the right by the projection $|\mathring{\Sigma}|$ we find that the conditions (A2) and (A3) must still hold so that λ remains exactly the same. Moreover, multiplying from the right by $\Gamma P_\alpha^\top$ we get with $P_\alpha P_\alpha^\top = I_m$ and $\Gamma\Gamma = P_\alpha^\top P_\alpha$ after some rearrangement the inequality

$$(\lambda - b)^\top \Gamma P_\alpha^\top = \lambda^\top \mathring{L} P_\alpha^\top - \mu^\top \leq \lambda^\top \mathring{L} P_\alpha^\top \ . \tag{A6}$$

Now the key observation is that this condition is linear in Γ and is strongest for the choice $\gamma_i = \text{sgn}(\lambda_i - b_i)$ for $i \in \alpha$ yielding the inequalities

$$|\lambda_i - b_i| \leq e_i^\top \mathring{L}^\top \lambda \quad \text{for} \quad i \in \alpha \ . \tag{A7}$$

In other words, $\mathring{x}$ is a solution of the branch problems (A4) if and only if it is for the worst case where $\gamma_i = \mathrm{sgn}(\lambda_i - b_i)$ for $i \in \alpha$. When coincidentally $\lambda_i = b_i$ we can define γ_i arbitrarily. Note that the complementarity condition $\mu^\top P_\alpha z(\mathring{x}) = 0$ associated with Equation (A4) is automatically satisfied at $\mathring{x}$ for any μ, since $P_\alpha \mathring{z} = 0$ by definition of the active index set α. These observations yield immediately:

Proposition A1 (Necessary and sufficient minimality condition). *Assume LIKQ holds in that $P_\alpha \mathring{Z}$ has full row rank $m = |\alpha|$. Then the point $\mathring{x}$ is a local minimizer of f if and only if we have tangential stationarity in that $a + \mathring{Z}^\top b$ belongs to the range of $\mathring{Z}^\top P_\alpha^\top$ and normal growth holds in that $|P_\alpha(\lambda - b)| \leq P_\alpha \mathring{L}^\top \lambda$.*

The verification that LIKQ holds and subsequently the test whether tangential stationarity is satisfied can be based on a QR decomposition of the active Jacobian $P_\alpha \mathring{Z} \in \mathbb{R}^{m \times n}$. The main expense here is the calculation of $\mathring{Z}$ itself, which requires one forward substitution on $(I - M - L\mathring{\Sigma})$ for each of n columns of Z and hence at most $ns^2/2$ fused multiply adds. Very likely this effort will already be made by any kind of active set method for reaching the candidate point $\mathring{x}$. Once the multiplier vector λ is obtained the remaining test (A7) for normal growth is almost for free so that we have a polynomial minimality criterion provided LIKQ holds. Otherwise one may assume a weaker generalization of the Mangasarian Fromovitz constrained qualification called MFKQ in [25]. However, we have conjectured in [19] that verifying MFKQ is probably already NP-hard.

Corollary A1 (Descent direction in the nonoptimal case). *Suppose that LIKQ holds. If tangential stationarity is violated there exits some direction $d \in \mathbb{R}^n$ such that $P_\alpha \mathring{Z} d = 0$ but $(a^\top + b^\top \mathring{Z})d < 0$, which implies descent in that $f(\mathring{x} + \tau d) < f(\mathring{x})$ for $\tau \gtrsim 0$. If tangential stationarity holds but normal growth fails there exists at least one $i \in \alpha$ with $|\lambda_i - b_i| > e_i^\top \mathring{L}^\top \lambda$. Defining $\gamma = \mathrm{sgn}(\lambda_i - b_i) e_i \in \mathbb{R}^s$, any d satisfying $P_\alpha (I - \mathring{L}\Gamma)^{-1} \mathring{Z} d = P_\alpha \gamma$ is a descent direction.*

Proof. In the first case it is clear that $\mathring{x} + \tau d \in \mathcal{P}_{\mathring{\sigma}}$ for $\tau \gtrsim 0$ since the components of $z(\mathring{x} + \tau d)$ with indices in α stay zero and the others vary only slightly. Then the directional derivative of $f(.)$ at $\mathring{x}$ in direction τd is given by

$$\tau a^\top d + \tau b^\top \mathring{Z} d = \tau(a^\top d + b^\top \mathring{Z} d) < 0,$$

which proves the first assertion. Otherwise, λ is well defined and we can choose $i \in \alpha$ with $|\lambda_i - b_i| > e_i^\top \mathring{L}^\top \lambda$. Setting $\gamma = \gamma_i e_i$ with $\gamma_i = \mathrm{sgn}(\lambda_i - b_i) e_i$, one obtains for d with $P_\alpha(I - \mathring{L}\Gamma)^{-1} \mathring{Z} d = \gamma$ that $\mathring{x} + \tau d \in \mathcal{P}_{\mathring{\sigma} + \gamma}$ for $\tau \gtrsim 0$. On that polyhedron the Lagrange multiplier vector μ is also well defined by Equation (A6) but we have

$$\mu_i = e_i^\top \mathring{L}^\top \lambda - (\lambda_i - b_i)\gamma_i = e_i^\top \mathring{L}^\top \lambda - |\lambda_i - b_i| < 0.$$

Then we get the directional derivative of $f(.)$ at $\mathring{x}$ in direction τd

$$\tau a^\top d + \tau b^\top (I - \mathring{L}\Gamma)^{-1} \mathring{Z} d = \tau(-\lambda^\top \mathring{Z} d + \lambda^\top \mathring{Z} d + \mu^\top P_\alpha \Gamma (I - \mathring{L}\Gamma)^{-1} \mathring{Z} d)$$
$$= \tau \mu_i \gamma_i^2 < 0,$$

where we have used identity (A5). Hence we have again descent, which completes the proof. □

Corollary A2 (Optimality via Reflection). *Suppose an $\mathring{x}$ where LIKQ holds has been reached by minimizing $\check{f}(x) + \hat{g}^\top x$ with $\hat{g} = \nabla \hat{f}_\sigma$ for $0 \notin \sigma > \mathring{\sigma}$. Then $\mathring{x}$ is a local minimizer of f on $\mathbb{R}^n$ if and only if it is also a minimizer of $\check{f}(x) + \nabla \hat{f}_{\tilde{\sigma}}^\top x$ with $\tilde{\sigma} = \sigma \triangleright \mathring{\sigma}$ as defined in (15).*

Proof. By assumption $\mathring{x}$ solves one of the branch problems of f itself. Hence we must have tangential stationarity (A5) with the corresponding $\Gamma = \mathrm{diag}(\gamma)$ for $\gamma = \sigma - \mathring{\sigma}$. Since $\tilde{\sigma} - \mathring{\sigma} = -\gamma$ we conclude from (A6) that

$$(\lambda - b)^\top \Gamma P_\alpha^\top \leq \lambda^\top \mathring{L} P_\alpha^\top \geq (\lambda - b)^\top (-\Gamma) P_\alpha^\top = -(\lambda - b)^\top \Gamma P_\alpha^\top$$

which implies that
$$\left|(\lambda - b)^\top P_\alpha^\top\right| = \left|(\lambda - b)^\top \Gamma P_\alpha^\top\right| \leq \lambda^\top \mathring{L} P_\alpha^\top. \tag{A8}$$

Hence both tangential stationarity and normal growth are satisfied, which completes the proof by Proposition A1 as the converse implication is trivial. □

The key conclusion is that if an $\mathring{x}$ is the solution of two complementary convex problems it must be locally optimal in the full dimensional space $\mathbb{R}^n$. Hence one can establish local optimality just using the preferred convex solver. If this test fails one naturally obtains descent to function values below $f(\mathring{x})$ until eventually a local minimizer is found.

Appendix A.1. Equivalence to DC Optimality Condition

Using the explicit expressions given in Lemma 1 we find that (see [18])
$$\partial^L f(\mathring{x}) = \bigcup_{0 = \gamma^\top \mathring{\sigma}} \left\{ a^\top + b^\top (I - \mathring{L}\Gamma)^{-1} \mathring{Z} \right\}, \tag{A9}$$

where γ ranges over all complements of $\mathring{\sigma}$ such that $\mathring{\sigma} + \gamma \in \{-1,1\}^s$ is definite. Similarly we obtain with
$$\tilde{b}^\top \equiv |b|^\top (I - |M| - 2|L|)^{-1} |L| \geq 0 \in \mathbb{R}^s$$

the limiting differentials of the convex and the concave part as
$$\partial^L \check{f}(\mathring{x}) = \bigcup_{0 = \gamma^\top \mathring{\sigma}} \left\{ a^\top + (b^\top + \tilde{b}^\top \mathring{\Sigma} + \tilde{b}^\top \Gamma)(I - \mathring{L}\Gamma)^{-1} \mathring{Z} \right\}, \tag{A10}$$
$$\partial^L \hat{f}(\mathring{x}) = \bigcup_{0 = \gamma^\top \mathring{\sigma}} \left\{ a^\top + (b^\top - \tilde{b}^\top \mathring{\Sigma} - \tilde{b}^\top \Gamma)(I - \mathring{L}\Gamma)^{-1} \mathring{Z} \right\}. \tag{A11}$$

Hence we have an explicit representation for the limiting gradients of f as well as its convex and concave part $\check{f}$ and $\hat{f}$ at $\mathring{x}$. It is easy to see that the minimality condition (A5) requires a to be in the range of $\mathring{Z}^\top$ so that we have again $a^\top = -\lambda^\top \mathring{Z}$ yielding
$$\partial^L \check{f}(\mathring{x}) = \bigcup_{0 = \gamma^\top \mathring{\sigma}} \left\{ (b^\top - \lambda^\top + \lambda^\top \mathring{L}\Gamma + \tilde{b}^\top \mathring{\Sigma} + \tilde{b}^\top \Gamma)(I - \mathring{L}\Gamma)^{-1} \mathring{Z} \right\}, \tag{A12}$$
$$\partial^L \hat{f}(\mathring{x}) = \bigcup_{0 = \gamma^\top \mathring{\sigma}} \left\{ (b^\top - \lambda^\top + \lambda^\top \mathring{L}\Gamma - \tilde{b}^\top \mathring{\Sigma} - \tilde{b}^\top \Gamma)(I - \mathring{L}\Gamma)^{-1} \mathring{Z} \right\}. \tag{A13}$$

We had hoped to be able to derive directly from these expressions that normal growth implies the condition (39), but we have so far not been able to do so. However, we can indirectly derive the following equivalence.

Corollary A3 (First order minimality condition). *Under LIKQ the limiting differential $\partial^L \hat{f}(\mathring{x})$ is contained in the convex hull of $-\partial^L \check{f}(\mathring{x})$ if and only if tangential stationarity and normal growth condition hold according to Proposition A1.*

References

1. Joki, K.; Bagirov, A.; Karmitsa, N.; Mäkelä, M. A proximal bundle method for nonsmooth DC optimization utilizing nonconvex cutting planes. *J. Glob. Optim.* **2017**, *68*, 501–535. [CrossRef]
2. Tuy, H. DC optimization: Theory, methods and algorithms. In *Handbook of Global Optimization*; Springer: Boston, MA, USA, 1995; pp. 149–216.
3. Rump, S. Fast and parallel interval arithmetic. *BIT* **1999**, *39*, 534–554. [CrossRef]
4. Bačák, M.; Borwein, J. On difference convexity of locally Lipschitz functions. *Optimization* **2011**, *60*, 961–978.

5. Griewank, A.; Walther, A. *Evaluating Derivatives: Principles and Techniques of Algorithmic Differentiation*; Society for Industrial and Applied Mathematics: Philadelphia, PA, USA , 2008.
6. Gürbüzbalaban, M.; Overton, M. On Nesterov's nonsmooth Chebyshev-Rosenbrock functions. *Nonlinear Anal. Theory Methods Appl.* **2012**, *75*, 1282–1289.
7. Griewank, A.; Walther, A. First and second order optimality conditions for piecewise smooth objective functions. *Optim. Methods Softw.* **2016**, *31*, 904–930. [CrossRef]
8. Strekalovsky, A. Local Search for Nonsmooth DC Optimization with DC Equality and Inequality Constraints. In *Numerical Nonsmooth Optimization. State of the Art Algorithms*; Springer Nature Switzerland AG: Cham, Switzerland, 2020; pp. 229–261.
9. Hansen, E. (Ed.) The centred form. In *Topics in Interval Analysis*; Oxford University Press: Oxford, UK , 1969; pp. 102–105.
10. Scholtes, S. *Introduction to Piecewise Differentiable Functions*; Springer: New York, NY, USA, 2012.
11. Griewank, A. On Stable Piecewise Linearization and Generalized Algorithmic Differentiation. *Optim. Methods Softw.* **2013**, *28*, 1139–1178. [CrossRef]
12. Griewank, A.; Bernt, J.U.; Radons, M.; Streubel, T. Solving piecewise linear equations in abs-normal form. *Linear Algebra Appl.* **2015**, *471*, 500–530. [CrossRef]
13. Griewank, A.; Walther, A.; Fiege, S.; Bosse, T. On Lipschitz optimization based on gray-box piecewise linearization. *Math. Program. Ser. A* **2016**, *158*, 383–415. [CrossRef]
14. Golub, G.; Van Loan, C. *Matrix Computations*, 4th ed.; Johns Hopkins University Press: Baltimore, MD, USA, 2013.
15. Rockafellar, R.; Wets, R.B. *Variational Analysis*; Springer: Berlin/Heidelberg, Germany, 1998.
16. Fukuda, K.; Gärtner, B.; Szedlák, M. Combinatorial redundancy detection. *Ann. Oper. Res.* **2018**, *265*, 47–65. [CrossRef]
17. Le Thi, H.; Pham Dinh, T. DC programming and DCA: Thirty years of developments. *Math. Program. Ser. B* **2018**, *169*, 5–68. [CrossRef]
18. Griewank, A.; Walther, A. Beyond the Oracle: Opportunities of Piecewise Differentiation. In *Numerical Nonsmooth Optimization. State of the Art Algorithms*; Springer: Cham, Switzerland, 2020; pp. 331–361.
19. Walther, A.; Griewank, A. Characterizing and testing subdifferential regularity in piecewise smooth optimization. *SIAM J. Optim.* **2019**, *29*, 1473–1501. [CrossRef]
20. Griewank, A. Who invented the reverse mode of differentiation? *Doc. Math.* **2012**, pp. 389–400.
21. Linnainmaa, S. Taylor expansion of the accumulated rounding error. *BIT* **1976**, *16*, 146–160. [CrossRef]
22. Sun, W.; Sampaio, R.; Candido, M. Proximal point algorithm for minimization of DC function. *J. Comput. Math.* **2003**, *21*, 451–462.
23. Griewank, A.; Walther, A. Finite convergence of an active signature method to local minima of piecewise linear functions. *Optim. Methods Softw.* **2019**, *34*, 1035–1055. [CrossRef]
24. Griewank, A.; Walther, A. *The True Steepest Descent Methods Revisited*; Technical Report; Humboldt-Universität zu Berlin: Berlin, Germany, 2020.
25. Griewank, A.; Walther, A. Relaxing kink qualifications and proving convergence rates in piecewise smooth optimization. *SIAM J. Optim.* **2019**, *29*, 262–289. [CrossRef]
26. Niu, Y. Programmation DC & DCA en Optimisation Combinatoire et Optimisation Polynomiale via les Techniques de SDP. Ph.D. Thesis, INSA Rouen, Rouen, France, 2010.
27. Fiege, S.; Walther, A.; Kulshreshtha, K.; Griewank, A. Algorithmic differentiation for piecewise smooth functions: A case study for robust optimization. *Optim. Methods Softw.* **2018**, *33*, 1073–1088. [CrossRef]
28. Walther, A.; Griewank, A. Getting Started with ADOL-C. In *Combinatorial Scientific Computing*; Chapman & Hall/CRC Computational Science Series; CRC Press: Boca Raton, FL, USA, 2012; pp. 181–202.
29. Hascoët, L.; Pascual, V. The Tapenade Automatic Differentiation tool: Principles, Model, and Specification. *ACM Trans. Math. Softw.* **2013**, *39*, 20:1–20:43.

© 2020 by the authors. Licensee MDPI, Basel, Switzerland. This article is an open access article distributed under the terms and conditions of the Creative Commons Attribution (CC BY) license (http://creativecommons.org/licenses/by/4.0/).

Article

On the Use of Biased-Randomized Algorithms for Solving Non-Smooth Optimization Problems

Angel Alejandro Juan [1,*], Canan Gunes Corlu [2], Rafael David Tordecilla [3] and Rocio de la Torre [4] and Albert Ferrer [5]

[1] Internet Interdisciplinary Institute (IN3), Department of Computer Science, Multimedia and Telecommunication, Universitat Oberta de Catalunya & Euncet Business School, 08018 Barcelona, Spain
[2] Metropolitan College, Boston University, Boston, MA 02215, USA; canan@bu.edu
[3] Internet Interdisciplinary Institute (IN3), Universitat Oberta de Catalunya & Universidad de La Sabana, 08018 Barcelona, Spain; rtordecilla@uoc.edu
[4] Institute for Advanced Research in Business and Economics (INARBE), Business Administration Department, Public University of Navarre, 31006 Pamplona, Spain; rocio.delatorre@unavarra.es
[5] Department of Applied Mathematics, Universitat Politecnica de Catalunya, 08028 Barcelona, Spain; alberto.ferrer@upc.edu
* Correspondence: ajuanp@uoc.edu; Tel.: +34-932-53-2300

Received: 13 December 2019; Accepted: 23 December 2019; Published: 25 December 2019

Abstract: Soft constraints are quite common in real-life applications. For example, in freight transportation, the fleet size can be enlarged by outsourcing part of the distribution service and some deliveries to customers can be postponed as well; in inventory management, it is possible to consider stock-outs generated by unexpected demands; and in manufacturing processes and project management, it is frequent that some deadlines cannot be met due to delays in critical steps of the supply chain. However, capacity-, size-, and time-related limitations are included in many optimization problems as hard constraints, while it would be usually more realistic to consider them as soft ones, i.e., they can be violated to some extent by incurring a penalty cost. Most of the times, this penalty cost will be nonlinear and even noncontinuous, which might transform the objective function into a non-smooth one. Despite its many practical applications, non-smooth optimization problems are quite challenging, especially when the underlying optimization problem is *NP-hard* in nature. In this paper, we propose the use of biased-randomized algorithms as an effective methodology to cope with *NP-hard* and non-smooth optimization problems in many practical applications. Biased-randomized algorithms extend constructive heuristics by introducing a nonuniform randomization pattern into them. Hence, they can be used to explore promising areas of the solution space without the limitations of gradient-based approaches, which assume the existence of smooth objective functions. Moreover, biased-randomized algorithms can be easily parallelized, thus employing short computing times while exploring a large number of promising regions. This paper discusses these concepts in detail, reviews existing work in different application areas, and highlights current trends and open research lines.

Keywords: non-smooth optimization; biased-randomized algorithms; heuristics; soft constraints

1. Introduction

Optimization models are used in many practical situations to represent decision-making challenges in areas such as computational finance, transportation and logistics, telecommunication networks, smart cities, etc. [1]. Many of these challenges can be transformed into optimization problems (OPs) that can be then solved using a plethora of methods of both exact and approximate nature. Typically, solving an OP implies exploring a vast solution space while searching for one solution that minimizes or maximizes a given objective function. In addition, the solution has to satisfy a series of constraints in order to be a feasible one [2]. It is frequent to model these OPs by using linear programming (LP), integer programming (IP), or mixed integer linear programming (MILP) methods. Unfortunately, in many real-life situations, these OPs are also *NP-hard*, which implies that the computing time requested to find an optimal solution grows extraordinarily fast as the size of the problem increases [3]. Hence, one has to make use of heuristic-based algorithms if reasonably good solutions are needed in short computing times for large-scale *NP-hard* OPs [4]. Moreover, the effective use of exact methods might be also limited whenever the mathematical model does not comply with desirable properties such as convexity or smoothness. In particular, the existence of non-convex and non-smooth objective functions might limit the efficiency of gradient-based optimization methods.

Bagirov and Yearwood [5] discuss a non-smooth OP called the minimum sum-of-squares clustering problem. According to the authors, previously employed approaches such as dynamic programming, branch-and-bound, or the *k*-means algorithm are efficient only for small instances of this problem. The authors also support the idea that the use of heuristic-based approaches becomes necessary for large-size instances. Similarly, Bagirov et al. [6] analyze another non-smooth OP related to the facility location problem in a wireless sensor network. Roy et al. [7] study non-smooth power-flow problems, while Lu et al. [8] propose an adaptive hybrid differential evolution algorithm to cope with a non-smooth version of the dynamic economic dispatch problem. To the best of our knowledge, however, there is a lack of publications considering realistic non-smooth cost functions in many OPs. Nevertheless, OPs with soft constraints might frequently appear in real-life applications. As discussed in Hashimoto et al. [9], "in real-world simulations, time windows and capacity constraints can be often violated to some extent". Hence, for example, in cost minimization, problems violating these soft constraints might generate penalty costs that might be taken into account in the objective function. These penalty costs will typically come in the form of a piecewise cost function, which can transform the objective function into a non-smooth one.

This paper reviews different examples of OPs with non-smooth objective functions and then analyzes how biased-randomized algorithms (BRAs) can constitute an effective methodology to generate reasonably good solutions in very short computing times. As described in Ferone et al. [10], BRAs make use of skewed probability distributions to integrate a "biased" (nonuniform) random behavior into a heuristic. This allows one to quickly generate a large set of alternative good solutions by simply changing the seed of the pseudo-random number generator [11,12]. Hence, each execution of the BRA can be seen as an individual "agent" searching the solution space following the logic behind the heuristic but starting from a different point and using a different searching (Figure 1). Moreover, the execution of these BRA agents can be performed in parallel, thus consuming virtually the same time as the original heuristic (i.e., milliseconds in most cases).

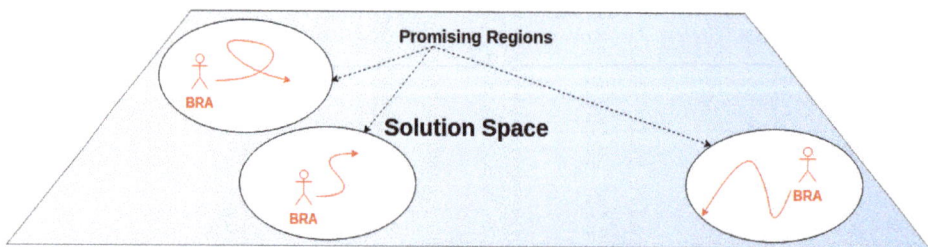

Figure 1. Exploring the solution space using biased-randomization algorithms.

According to our previous experience with using BRAs to solve OPs in different application fields, these algorithms can be especially useful in cases where the solution space is highly irregular (non-convex and/or non-smooth) and requires an extensive exploration stage, thus reducing the effectiveness of traditional optimization methods. Actually, BRAs have been already proposed to solve non-smooth OPs in different application areas. For instance, they have been used to solve different rich and realistic variants of the well-known vehicle routing problem (VRP), including the two-dimensional VRP [13], VRP variants with horizontal cooperation [14], multi-agent versions of the VRP [15], the location routing problem [16], the fleet mixed VRP with backhauls [17,18], the multi-period VRP [19], and even other versions of the multi-depot VRP [20]. BRAs have also been employed in solving other OPs, such as the single-round divisible load scheduling [21], the stochastic flow-shop scheduling [22], scheduling heterogeneous multi-round systems [23], the minimization of open stacks problem [24], the dynamic home service routing [25], waste collection management [26], or the maximum quasi-clique problem [27].

Accordingly, the main contributions of this paper are as follows: (i) a discussion on the importance of considering non-smooth objective functions in realistic combinatorial OPs, mainly due to the existence of soft constraints which might be violated to some extent by incurring non-smooth penalty costs, and (ii) a discussion on how BRAs can be employed in different applications to solve these non-smooth OPs in short computing times. The remainder of the paper is structured as follows: Section 2 reviews some basic concepts related to non-smooth OPs. Section 3 presents a review of recent works on BRAs. Sections 4–6 review applications of BRAs to non-smooth OPs in logistics, transportation, and scheduling, respectively. Section 7 provides an overview on current trends and open research lines. Finally, Section 8 concludes by highlighting the main contributions of this work.

2. Non-Smooth Optimization Problems

OPs can be broadly classified as convex or non-convex. Convex OPs are usually characterized by a convex objective function and a set of constraints that form a convex region. Each constraint restricts the solution space to a convex region, and the intersection of these regions, which form the feasible solutions, is also convex. The main feature that makes convex OPs easy to work with is that any local optimum is also a global optimum. This significantly reduces the computational time yielding exact solutions in reasonable times. Therefore, if doable, it is of interest to convert any optimization problem into a convex OP. Despite the specific structure needed for a convex OP, we find several applications of it in real-life problems. For example, those problems that can be modeled as a linear programming model are convex problems because all linear functions are by definition convex [28]. Nevertheless, there are some other problems that cannot be modeled as a convex OP. Non-convex problems have either non-convex objective functions or non-convex feasible regions (or both). This brings in several challenges to solve these problems. The main challenge is that the solution methods employed for convex OPs cannot be directly applied for non-convex ones because of the availability of many disjoint regions in the solution space, each of which usually has its own local optimum. Therefore, it is easy for the algorithm to get trapped into one of these local optima, which may indeed be far away from the global optimum. Also, it is usually time-consuming—or even impossible—to demonstrate that the algorithm reached the global optimum or whether a feasible solution can be obtained.

Another way of classifying an OP is by whether it is a smooth or a non-smooth one. Smooth optimization problems have smooth objective functions and constraints. A smooth function has derivatives of all orders and is differentiable. On the contrary, a non-smooth one has an objective function—or at least one constraint—that does not possess at least one of the properties of a smooth function. Figure 2 shows an example of a one-dimensional function which is neither smooth nor convex.

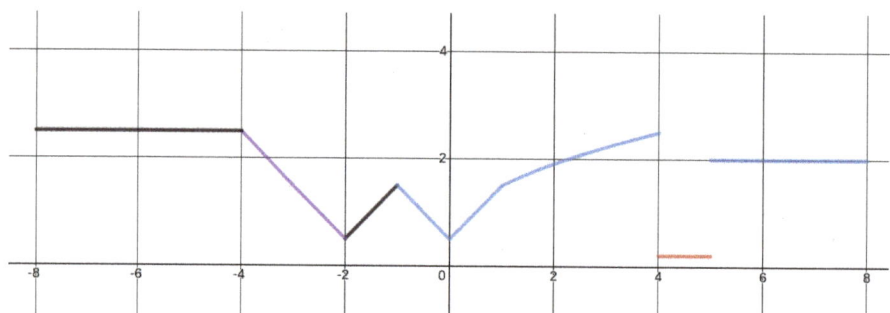

Figure 2. Example of a non-convex and non-smooth piecewise objective function.

From a combinatorial point of view, non-smooth OPs possess similar properties as non-convex OPs because they are time consuming to solve. The lack of derivative information makes it almost impossible to determine the direction in which the function is increasing or decreasing. Likewise, the solution space may also have several disjoint regions, each of which has its own local optimum. Unfortunately, non-convex and non-smooth OPs arise in several application domains, including telecommunication networks, economic load dispatch, portfolio optimization, vehicle routing, regression, or clustering problems. For instance, the minimum sum-of-squares clustering problem is solved by Bagirov et al. [29] and by Karmitsa et al. [30]. Both papers formulate the clustering problem as a non-smooth and non-convex optimization problem and make use of incremental algorithms. However, the former is based on the difference of convex functions and the latter is based on the limited memory bundle method. Real world data sets are used to test both approaches, demonstrating numerically their efficiency compared to other incremental algorithms. Difference of convex functions are also used by Bagirov et al. [31] to solve the nonparametric regression estimation problem. These authors propose an algorithm to minimize a non-convex and non-smooth empirical L_2-risk function. Synthetic and real-world data sets are used to test it. Compared to other algorithms, this approach is proved to be a good alternative in terms of computational time and several prediction indicators.

Several studies have investigated the applicability of well-known metaheuristic approaches—such as tabu search, artificial bee colony optimization, or particle swarm optimization—to solve non-smooth and non-convex OPs [32]. For example, tabu search has been used in Al-Sultan [33] for the clustering problem and in Oonsivilai et al. [34] for a telecommunication network problem. Ant colony optimization has been used to solve the non-smooth economic load dispatch problem in Hemamalini and Simon [35], while particle swarm optimization has been investigated for the same problem in Niknam et al. [36] and Basu [37]. Both ant colony optimization and particle swarm optimization have been utilized for the non-smooth portfolio selection problem in Schlüter et al. [38] and Corazza et al. [39], respectively. The remainder of this paper discusses the use of BRAs in solving non-smooth optimization problems in logistics, transportation, and scheduling.

3. Basic Concepts on Biased-Randomized Algorithms

Pure greedy constructive heuristics are algorithms that iteratively build a solution by selecting the next movement from a list of candidates. Such candidates have been sorted previously according to some criteria, such as costs, savings, profits, etc. These heuristics typically select the "most promising" (in the short run) candidate from the list. Since they follow a constructive logic, a good final solution is expected by the end of the procedure. Nevertheless, these algorithms are deterministic, i.e., the solution is always the same every time the heuristic is executed. This means that the exploration process is poor, which prevents the algorithm from finding better solutions unless more complex searching structures—i.e., local searches and perturbation movements—are considered by investing more computing time. Examples of such heuristics are the well-known savings heuristic for the VRP [40], the nearest neighbor criterion for the traveling salesman problem [41], or the shortest processing time dispatching rule for some scheduling problems [42].

As described in Juan et al. [43], using a skewed (nonuniform) probability distribution to introduce a biased-randomization behavior into the process that selects the candidates from the sorted list is an efficient way of generating better solutions. The idea is to assign a weighted probability to each candidate in the list, in such a way that the more promising candidates—those at the top of the list—receive a higher probability of being selected than those below them. This randomization process leads to the generation of slightly different solutions every time the algorithm is executed. Hence, multiple executions of a BRA—either completed in a sequential or in a parallel mode—will yield a set of alternative solutions, all of them based on the logic behind the heuristic. Since we are executing many biased-random variations of the constructive procedure defined by the heuristic, chances are that some of these "near-greedy" heuristics lead to solutions that outperform the one generated by the greedy heuristic [10]. Algorithm 1 shows a pseudo-code description of a basic BRA that performs in a sequential way.

Algorithm 1: Biased-Randomized Algorithm (BRA; basic sequential version).

---
1 bestSol ← *execute the deterministic (greedy) heuristic*
2 **while** *more time is allowed* **do**
3   seed ← *select a seed for the pseudo-random number generator*
4   rng ← *start a pseudo-random number generator with seed*
5   dist ← *select a skewed probability distribution (and its parameters)*
6   newSol ← *start an empty solution*
7   list ← *create the list of 'building blocks' for the solution*
8   sortedList ← *sort the list according to the heuristic logic*
9   **while** *sortedList is not empty* **do**
10   nextElement ← *use dist and rng to select and extract the next block from sortedList*
11   **if** *nextElement can be added without losing feasibility* **then**
12   newSol ← *add nextElement to the incumbent solution*
13   **end**
14   **end**
15 **end**
16 **if** *newSol is better than bestSol* **then**
17   bestSol ← newSol
18 **end**
19 **return** bestSol

Notice that, by using this approach, a broad exploration of the solution space is carried out, which might be specially beneficial in the case of highly irregular objective functions as the ones characterizing non-smooth OPs. The proposed methodology can be seen as a natural extension of the basic greedy randomized adaptive search procedure (GRASP) [44], as analyzed in Ferone et al. [10]. Instead of employing empirical probability distributions—which require time-consuming parameter fine tuning and thus might slow down computations—a theoretical probability distribution such as the geometric distribution or the decreasing triangular distribution can be used. Random variates from these theoretical distributions can be quickly generated by employing analytical expressions. Moreover, they tend to have less parameters and these are typically easy to set. Application fields such as food logistics [45], flow-shop scheduling [46], or mobile cloud computing [47] have successfully utilized geometric distributions to introduce biased-randomized processes during the selection of the candidates that are employed to construct a feasible solution. Figure 3 illustrates how geometric probability distributions with four different parameter values ($p \in \{0.1, 0.3, 0.6, 0.9\}$) will have a different behavior while assigning probabilities of being selected to the elements of the sorted list during the iterative construction of a biased-randomized solution.

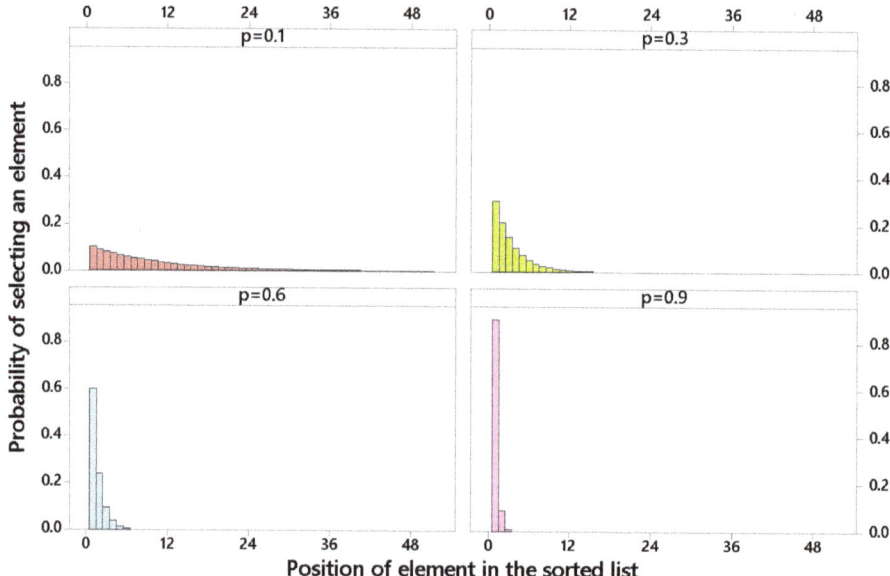

Figure 3. Biased-random sampling of elements from a list using a geometric distribution.

Thus, while for $p = 0.1$ the distribution is closer to a uniform one (i.e., the probabilities are distributed among a relatively large number of top positions in the sorted list), for $p = 0.9$, the behavior is closer to the greedy one that characterizes a classical heuristic, with the top element in the sorted list accumulating most of the chances of being the next selected element. Both extremes ($p \to 0$ and $p \to 1$) represent diversification and greediness, respectively. Usually, parameter values in the middle of both extremes are able to provide a better trade-off between these two cases, thus promoting some degree of diversification without losing the rational (domain-specific) criterion employed to sort the list.

4. Applications in Logistics

The field of logistics encompasses several problems, including supply chain design, facility location, warehouse management, etc. All of these problems have been studied extensively in the literature, mostly with the consideration of hard constraints and smooth objective functions. Nevertheless, as previously discussed, real-world problems in the field of logistics may allow some constraints to be violated by incurring a penalty cost, which needs to be incorporated into the objective function. This typically leads to the emergence of non-smooth objective functions. Therefore, traditional exact methods cannot always be efficiently employed to solve these problems and heuristic-based algorithms are required. This section focuses on the use of BRAs in solving the facility location problem (FLP) [48] and its variants. This problem consists of locating a set of facilities—e.g., production plants, distribution centers, warehouses, etc.—from which a set of customers must be served. Basic decisions are as follows: (i) which potential facilities must be open (or remain open) and which ones must be closed (or not open) and (ii) how to allocate customers to open facilities. This problem is *NP-hard* [49]. Moreover, facilities can be considered capacitated or uncapacitated. The former refers to the case in which each facility has a limited capacity that cannot be exceeded by the total demand served from there. In the latter, the facilities' total capacity is virtually infinite or at least much greater than the cumulative demand of all customers.

BRAs have been applied successfully to solve both capacitated and uncapacitated FLPs. The latter has been tackled mainly considering hard constraints [50]. In Correia and Melo [51], the authors considered a multi-period FLP in which customers are sensitive to delivery lead times (i.e., some flexibility is allowed regarding the delivery dates). Using similar concepts, Estrada-Moreno et al. [52] consider soft constraints and a non-smooth and non-convex objective function for the single-source capacitated FLP. In this context, "single-source" refers to an additional constraint stating that each customer must be served from just one facility. The capacity of each facility may be exceeded by the consideration soft constraints. In real world, decision-makers manage this by using strategies such as storing safety stocks, performing emergency deliveries, and outsourcing part of the customers' service. These strategies tend to generate additional costs that need to be considered as well during the optimization process. The aforementioned authors propose the following model to represent the single-source FLP with soft capacity constraints:

$$\text{Minimize} \sum_{i \in J} f_i^* y_i + \sum_{(i,j) \in I \times J} c_{ij} x_{ij}$$

subject to:

$$\sum_{j \in J} x_{ij} = 1 \quad \forall i \in I$$

$$x_{ij} \leq y_j \quad \forall (i,j) \in I \times J$$

$$y_j(1 - y_j) = 0 \quad \forall j \in J$$

$$x_{ij}(1 - x_{ij}) = 0 \quad \forall (i,j) \in I \times J$$

$$y_j \in \mathbb{R} \quad \forall j \in J$$

$$x_{ij} \in \mathbb{R} \quad \forall (i,j) \in I \times J$$

In this model, x_{ij} is a binary variable that takes the value of 1 if customer i is serviced by facility j (0 otherwise). Similarly, y_j is another binary variable that takes the value 1 if facility j is open; c_{ij} is the service cost of assigning customer i to facility j; and f_j^* is a piecewise function representing the cost of opening a facility j:

$$f_j^* = \begin{cases} f_j & \text{if } \sum_{i \in I} d_i x_{ij} \leq s_j y_j \\ f_j + \lambda \left(d_j^*, s_j \right) & \text{otherwise} \end{cases}$$

where $d_i > 0$ is the demand of customer i; $s_j \gg \max\{d_i\}$ is the nominal capacity of facility j; $d_j^* = \sum_{i \in I} d_i x_{ij}$ is defined for any $j \in J$; and $\lambda \left(d_j^*, s_j \right)$ is a non-smooth function which will be applied whenever the total demand assigned to facility j exceeds its maximum capacity s_j.

A BRA is integrated within an iterated local search metaheuristic to solve the OP above. The algorithm contains the following components: (i) an initial solution generation, based on a BRA; (ii) a local search procedure composed of functions that open or close facilities; (iii) a perturbation procedure that destroys the current solution by opening a number of closed facilities and reallocating all customers to the newly open facilities; and (iv) an acceptance criterion based on the concept of "credit", in which a solution with a worse cost is accepted if this credit is not exceeded. The objective is to explore other regions of the solution space to escape from local optima. A total of 60 small-, medium-, and large-scale instances are used to test this approach. Different levels of penalties are also tested. Authors demonstrate the advantages of using soft constraints, obtaining costs that are lower than the optimal ones found in the literature for hard constraints. Authors show that, if penalty costs are low or moderate, hard constraints' violation is worth because some facilities do not need to be open and a more efficient allocation of customers can be made. Finally, a comparison between their

BRA-inspired metaheuristic and the solution provided by the commercial tool *LocalSolver* is drawn. Both approaches obtain similar solutions for small- and medium-scale instances, but the BRA-based algorithm proves to be superior for large-scale instances.

5. Applications in Transportation

Transportation and distribution are two fields belonging to the operational level of decisions in logistics. The vehicle routing problem [53,54] and the arc-routing problem (ARP) [55,56] are two well-known optimization problems in the area of transportation and freight distribution. A traditional VRP consists of a graph formed by a set of nodes and a set of arcs. One of the nodes represents a depot and the rest represent customers, which are connected to each other by the set of arcs. A network of routes must be designed to visit each customer in order to meet a known demand. A single vehicle departing and returning to the depot is assigned to each route. The objective is to minimize the total cost of traversing the arcs. The traditional ARP is similar to the VRP, but the former assigns a demand to each arc and not to each node. Moreover, in the ARP, the underlying graph is not usually a complete one. These problems are *NP-hard*, i.e., as the number of customers grows, the quantity of alternative solutions increases almost exponentially. Therefore, heuristic and metaheuristic algorithms have been employed extensively to solve these OPs. The problems become even more difficult to solve in the presence of non-smooth and non-convex objective functions, as when soft constraints are considered. In this context, soft constraints would include time window constraints or capacity constraints that are allowed to be violated to some extent [9]. These soft constraints also allow decision-makers to consider more realistic models that take into account different management strategies and policies. For instance, customers would accept a delayed delivery if the supplier offers a discount. Likewise, a percentage of the deliveries can be outsourced if in-house capacity is exceeded.

In the VRP case, Juan et al. [43] consider a capacitated version of the problem with a non-smooth and non-convex objective function and soft constraints. These authors propose a BRA-based approach called *MIRHA*. This is a multi-start procedure consisting of two phases: a first phase in which a biased-randomized version of a constructive heuristic is designed according to a geometric probability distribution and a second (improvement) phase in which an adaptive local search procedure is implemented. Several instances from the literature are used to test the proposed algorithm and to compare it with a traditional GRASP. In general, the new algorithm outperforms the existing ones in terms of solution quality (efficiency), both in the presence of hard and soft constraints. In the case of the ARP, De Armas et al. [57] propose a BRA to solve the capacitated version of the problem with a non-smooth and non-convex objective function. The base heuristic considered is *SHARP* [58]. Firstly, they propose the following model:

$$\text{Minimize} \sum_{\rho \in \mathcal{S}} c_\rho^*$$

subject to:

$$\mathcal{S} \in CSR$$

$$\sum_{(i,j) \in \rho} q_{ij} x_{ij}^k \leq Q \quad \forall \rho \in \mathcal{S}, \forall k \in T$$

$$x_{ij}^k (1 - x_{ij}^k) = 0 \quad \forall (i,j) \in \rho, \forall \rho \in \mathcal{S}, \forall k \in T$$

where ρ represents a route in a set of routes $\mathcal{S}$; CSR represents a complete set of routes (i.e., a solution); c_ρ^* is the total cost of using route ρ; q_{ij} is the demand of arc (i,j); and x_{ij}^k is a binary variable that takes the value 1 if and only if the arc (i,j) is covered by a vehicle k in the set of vehicles T. The cost function associated with any route ρ is defined as a piecewise function as follows:

$$c_\rho^* = \begin{cases} c_\rho & \text{if } c_\rho \leq C \\ c_\rho + \lambda\left(c_\rho, C\right) & \text{otherwise} \end{cases} \qquad (1)$$

In the former expression, $\lambda\left(c_\rho, C\right)$ is a non-smooth function which will be applied whenever the actual route cost exceeds the threshold value allowed for any route, C. In this work, the associated penalty factor is linearized to obtain a truncated version of the problem. Then, the authors propose a BRA combined with an iterated local search metaheuristic. This hybrid algorithm is divided into four main phases: (i) an initial solution generation using a biased-randomized version of the *SHARP* heuristic; (ii) a perturbation procedure based on destruction-reconstruction strategies; (iii) a local search phase using cache memory; and (iv) the use of an acceptance criterion based on a simulated annealing procedure. A total of 87 artificial and real-world instances are used to test this approach. The mathematical model is solved using the *CPLEX* commercial tool and is used to obtain lower bounds for optimal costs. Thus, the performance of the metaheuristic is assessed, obtaining important average gap reductions regarding previous methods.

The results obtained in the previous studies demonstrate clear advantages of considering soft constraints over hard ones. Moreover, the associated models are better representation of the real-world problems. For example, budget limits established per route can be violated in the model as it is done in real life. The soft constraints are not free, but they enhance profitability. For instance, penalization costs must be incurred for violating budget limits. However, this violation leads to better route design, which generates savings for the company. In the end, it might be worthy to explore if the value of the savings compensates the penalties incurred. Finally, the consideration of soft constraints implies the construction of a more generic model, which includes a combination of soft and hard constraints. A more generic model yields more alternative solutions, and therefore, decision makers have more options to design a routing or distribution plan that better fits their utility function.

6. Applications in Scheduling

In the operations research field, scheduling OPs are among the most studied topics. According to Pinedo [59], "scheduling is a decision-making process [...] that deals with the allocation of resources to tasks over given time periods and its goal is to optimize one or more objectives." A typical example considers that the resources are machines, that the tasks are operations carried out by these machines, and that the objective is the minimization of the *makespan*—i.e., the completion time of the last task. This apparently simple definition actually includes a huge family of problems that are *NP-hard* as well.

BRAs have been proved to be useful also for scheduling problems. For instance, Martin et al. [15] used them in a multi-agent based framework to solve both routing and scheduling problems. Usually, hard constraints are considered in the literature. However, Ferrer et al. [60] solved the permutation flow-shop problem (PFSP) with a non-smooth objective function. This problem consists of a set of jobs that must be processed by a set of machines. Each job is composed of a set of operations in which the quantity is equal to the number of machines. Moreover, all operations in each job must be executed in the same sequence by the set of machines. The processing time of each operation in each machine is known, although it is different for each job. The idea is to determine the sequence in which jobs must be executed in order to minimize the makespan. One of the contributions of the aforementioned authors is the consideration of the failure-risk term in the objective function. This term is incorporated into the traditional makespan target. The failure-risk cost is incurred when a machine operates continuously without a break, which is highly usual when minimizing the makespan. This cost is equivalent to a penalty cost in logistics and transportation problems, and therefore, the failure-risk cost introduces a non-smooth component into the objective function. A mathematical model is proposed to tackle this problem. Then, a BRA is combined with an iterated local search to develop the solving approach. Its basic steps are as follows: (i) generation of an initial solution through a biased-randomized version of a classical heuristic; (ii) perturbation and local search procedures are implemented to improve the solution quality; and (iii) the consideration of an acceptance criterion, based on simulated annealing, which may accept worst solutions with the purpose of exploring the solution space and escape from local minima. A total of 120 benchmark instances are used to test this approach. Results show that explicit consideration of the reduction in failure-risk cost leads to a reduction of the total cost for all sets of instances (negative gaps are shown). The makespan cost increases, but the reduction in failure-risk costs compensate this rise. A comparison with other metaheuristic approaches shows that the performance of the new method is similar or even better. This novel approach proves to be useful for decision makers since they have more solution alternatives to select, given the particular goals of each company, i.e., for some decision-makers, the makespan may show a higher relevance, but for others, failure-risk cost may be a more important indicator.

7. General Insights from Previous Numerical Experiments

Based on the data provided by Juan et al. [43] for the non-smooth VRP, De Armas et al. [57] for the non-smooth ARP, Ferrer et al. [60] for the non-smooth permutation FSP, and Estrada-Moreno et al. [52] for the non-smooth FLP, Figure 4 shows percentage gaps between the corresponding BRA and the reference value employed in the corresponding work.

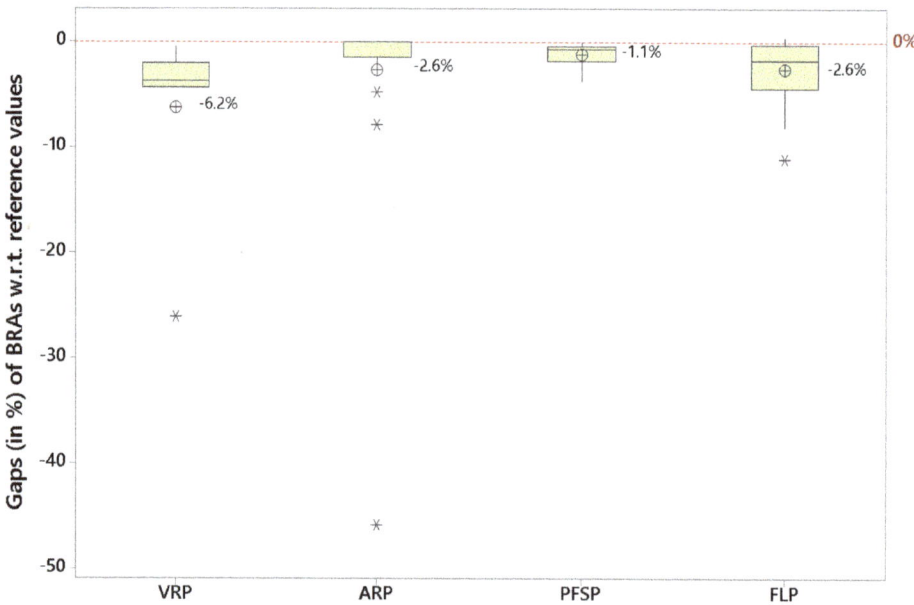

Figure 4. Percentage gaps between BRAs and reference values in non-smooth optimization problems (OPs).

Precaution has to be used while interpreting this figure, since these gaps depend on the particular OP being considered, the specific instances, the selected reference value, etc. However, some insights can be obtained: (i) in all four OPs, BRAs have been able to obtain negative gaps with respect the reference values, which in several cases represent the best-known solutions for the hard-constrained version of the problem; (ii) whenever soft constraints can be considered in a real-life scenario, it might pay off to design solutions that violate these constraints to some extent, since the associated benefits might overcome the corresponding penalties; (iii) since they combine good diversification (exploration) and heuristic-based rational searching of the solution space, BRAs constitute an effective tool to cope with non-smooth OPs with highly irregular objective functions; and (iv) in some particular cases, considering soft constraints instead of hard ones might generate noticeable improvements in the quality of the solution; hence, modeling and solving experts should always consider how really hard is a constraint in practice. Finally, it should be also noticed that, in all the four studies and regardless of the specific OP being solved and the application field, the authors illustrated with numerical examples the limitations of exact methods when solving non-smooth OPs. Despite this generalized conclusion, they also recommended to investigate combinations between exact methods (from global optimization and mathematical programming) and heuristic-based algorithms. The latter can play a more exploratory role and can identify promising areas, while the former can intensify the searching process inside these selected areas.

8. Conclusions

Many real-world problems can be more accurately modeled using soft constraints rather than hard ones. Soft constraints can be violated to some extent, and whenever this occurs, a penalty cost—which is usually defined via a piecewise function depending on the magnitude of the violation—has to be taken into account. Hence, non-smooth and non-convex optimization models are highly relevant in many practical applications. In general, decision makers may consider some constraints to be soft, specifically when the associated capacity limitations can be outsourced or certain delays in the service can be managed with the customer. This paper has reviewed several works in this area. These works refer to the consideration of non-smooth objective functions in popular OPs such as the vehicle routing problem, the arc routing problem, the facility location problem, and the permutation flow-shop problem. In all these cases, the use of BRAs has shown to be an effective tool to generate a myriad of high-quality solutions in short computing times, even for large-size instances of these *NP-hard* OPs. Also, these BRAs have outperformed other classical optimization methods from the areas of mathematical programming, global optimization, and even metaheuristics.

The reviewed papers demonstrate that using BRAs enhances the exploration of the solution space by generating iteratively. The execution of these algorithms might be easily parallelized by simply changing the seed of the pseudo-random number generator, which means that, in many cases, high-quality solutions—close to near-optimal ones—can be frequently obtained in real-time (less than a second). Part of the effectiveness of these algorithms lies in the fact that they preserve the logic of a good constructive heuristics while, at the same time, they offer a much larger exploration capability of the solution space. The paper has also discussed how these BRAs can be hybridized with classical metaheuristic frameworks—e.g., iterated local search, simulated annealing, etc.—in order to increase the searching process if more computing time is allowed.

Several research lines can be explored for future work: (i) the hybridization of the BRAs with the *ECAM* global optimization algorithm [61], so that the former can provide different starting points (exploration) that the latter can use to intensify the search in promising regions; (ii) other optimization problems can be considered as well, especially in application fields such as smart cities, e-commerce, computational finance, or bioinformatics; and (iii) considering even more realistic versions of the optimization problems by adding stochastic and dynamic conditions into them, for which hybridization of BRAs with simulation and machine learning techniques might be necessary.

Author Contributions: Conceptualization, A.A.J.; methodology, C.G.C., R.D.T., and R.d.l.T.; writing—original draft preparation, A.A.J., C.G.C., R.D.T., A.F., and R.d.l.T.; writing—review and editing, A.A.J., C.G.C., R.D.T., A.F., and R.d.l.T. All authors have read and agree to the published version of the manuscript.

Funding: This research was partially funded by the IoF2020 European project, AGAUR (2018-LLAV-00017), the Erasmus+ program (2018-1-ES01-KA103-049767), and the Spanish Ministry of Science, Innovation, and Universities (RED2018-102642-T).

Acknowledgments: We thank Dr. Napsu Karmitsa and Dr. Sona Taheri for inviting us to participate in this special issue in honor of Prof. Dr. Adil M. Bagirov.

Conflicts of Interest: The authors declare no conflict of interest.

References

1. Simpson, N.C.; Hancock, P.G. *Practical Operations Management*; Hercher: Naperville, IL, USA, 2013.
2. Papadimitriou, C.H.; Steiglitz, K. *Combinatorial Optimization: Algorithms and Complexity*; Prentice-Hall, Inc.: Upper Saddle River, NJ, USA, 1982.
3. Garey, M.R.; Johnson, D.S. *Computers and Intractability; A Guide to the Theory of NP-Completeness*; W. H. Freeman & Co.: New York, NY, USA, 1990.
4. Khamaru, K.; Wainwright, M.J. Convergence guarantees for a class of non-convex and non-smooth optimization problems. *J. Mach. Learn. Res.* **2019**, *20*, 1–52.
5. Bagirov, A.M.; Yearwood, J. A new nonsmooth optimization algorithm for minimum sum-of-squares clustering problems. *Eur. J. Oper. Res.* **2006**, *170*, 578–596. [CrossRef]
6. Bagirov, A.; Lai, D.T.H.; Palaniswami, M. A nonsmooth optimization approach to sensor network localization. In Proceedings of the 3rd International Conference on Intelligent Sensors, Sensor Networks and Information, Melbourne, Australia, 3–6 December 2007; pp. 727–732.
7. Roy, P.; Ghoshal, S.; Thakur, S. Biogeography based optimization for multi-constraint optimal power flow with emission and non-smooth cost function. *Expert Syst. Appl.* **2010**, *37*, 8221–8228. [CrossRef]
8. Lu, Y.; Zhou, J.; Qin, H.; Li, Y.; Zhang, Y. An adaptive hybrid differential evolution algorithm for dynamic economic dispatch with valve-point effects. *Expert Syst. Appl.* **2010**, *37*, 4842–4849. [CrossRef]
9. Hashimoto, H.; Ibaraki, T.; Imahori, S.; Yagiura, M. The vehicle routing problem with flexible time windows and traveling times. *Discret. Appl. Math.* **2006**, *154*, 2271–2290. [CrossRef]
10. Ferone, D.; Gruler, A.; Festa, P.; Juan, A.A. Enhancing and extending the classical GRASP framework with biased randomisation and simulation. *J. Oper. Res. Soc.* **2019**, *70*, 1362–1375. [CrossRef]
11. Faulin, J.; Gilibert, M.; Juan, A.A.; Vilajosana, X.; Ruiz, R. SR-1: A simulation-based algorithm for the capacitated vehicle routing problem. In Proceedings of the 2008 Winter Simulation Conference, Miami, FL, USA, 7–10 December 2008; pp. 2708–2716.
12. Juan, A.A.; Faulin, J.; Ruiz, R.; Barrios, B.; Gilibert, M.; Vilajosana, X. Using oriented random search to provide a set of alternative solutions to the capacitated vehicle routing problem. In *Operations Research and Cyber-Infrastructure*; Springer: Berlin/Heidelberg, Germany, 2009; pp. 331–345.
13. Domínguez Rivero, O.L.; Juan Pérez, A.A.; De La Nuez Pestana, I.A.; Ouelhadj, D. An ILS-biased randomization algorithm for the two-dimensional loading HFVRP with sequential loading and items rotation. *J. Oper. Res. Soc.* **2016**, *67*, 37–53. [CrossRef]
14. Quintero-Araujo, C.L.; Gruler, A.; Juan, A.A.; Faulin, J. Using horizontal cooperation concepts in integrated routing and facility-location decisions. *Int. Trans. Oper. Res.* **2019**, *26*, 551–576. [CrossRef]
15. Martin, S.; Ouelhadj, D.; Beullens, P.; Ozcan, E.; Juan, A.A.; Burke, E.K. A multi-agent based cooperative approach to scheduling and routing. *Eur. J. Oper. Res.* **2016**, *254*, 169–178. [CrossRef]
16. Quintero-Araujo, C.L.; Caballero-Villalobos, J.P.; Juan, A.A.; Montoya-Torres, J.R. A biased-randomized metaheuristic for the capacitated location routing problem. *Int. Trans. Oper. Res.* **2017**, *24*, 1079–1098. [CrossRef]
17. Belloso, J.; Juan, A.A.; Martinez, E.; Faulin, J. A biased-randomized metaheuristic for the vehicle routing problem with clustered and mixed backhauls. *Networks* **2017**, *69*, 241–255. [CrossRef]
18. Belloso, J.; Juan, A.A.; Faulin, J. An iterative biased-randomized heuristic for the fleet size and mix vehicle-routing problem with backhauls. *Int. Trans. Oper. Res.* **2019**, *26*, 289–301. [CrossRef]
19. Estrada-Moreno, A.; Savelsbergh, M.; Juan, A.A.; Panadero, J. Biased-randomized iterated local search for a multiperiod vehicle routing problem with price discounts for delivery flexibility. *Int. Trans. Oper. Res.* **2019**, *26*, 1293–1314. [CrossRef]
20. Calvet, L.; Ferrer, A.; Gomes, M.I.; Juan, A.A.; Masip, D. Combining statistical learning with metaheuristics for the multi-depot vehicle routing problem with market segmentation. *Comput. Ind. Eng.* **2016**, *94*, 93–104. [CrossRef]
21. Brandão, J.S.; Noronha, T.F.; Resende, M.G.C.; Ribeiro, C.C. A biased random-key genetic algorithm for single-round divisible load scheduling. *Int. Trans. Oper. Res.* **2015**, *22*, 823–839. [CrossRef]
22. Gonzalez-Neira, E.M.; Ferone, D.; Hatami, S.; Juan, A.A. A biased-randomized simheuristic for the distributed assembly permutation flowshop problem with stochastic processing times. *Simul. Model. Pract. Theory* **2017**, *79*, 23–36. [CrossRef]

23. Brandão, J.S.; Noronha, T.F.; Resende, M.G.C.; Ribeiro, C.C. A biased random-key genetic algorithm for scheduling heterogeneous multi-round systems. *Int. Trans. Oper. Res.* **2017**, *24*, 1061–1077. [CrossRef]
24. Gonçalves, J.F.; Resende, M.G.C.; Costa, M.D. A biased random-key genetic algorithm for the minimization of open stacks problem. *Int. Trans. Oper. Res.* **2016**, *23*, 25–46. [CrossRef]
25. Fikar, C.; Juan, A.A.; Martinez, E.; Hirsch, P. A discrete-event driven metaheuristic for dynamic home service routing with synchronised trip sharing. *Eur. J. Ind. Eng.* **2016**, *10*, 323–340. [CrossRef]
26. Gruler, A.; Fikar, C.; Juan, A.A.; Hirsch, P.; Contreras-Bolton, C. Supporting multi-depot and stochastic waste collection management in clustered urban areas via simulation–optimization. *J. Simul.* **2017**, *11*, 11–19. [CrossRef]
27. Pinto, B.Q.; Ribeiro, C.C.; Rosseti, I.; Plastino, A. A biased random-key genetic algorithm for the maximum quasi-clique problem. *Eur. J. Oper. Res.* **2018**, *271*, 849–865. [CrossRef]
28. Boyd, S.; Vandenberghe, L. *Convex Optimization;* Cambridge University Press: Cambridge, UK, 2004.
29. Bagirov, A.M.; Taheri, S.; Ugon, J. Nonsmooth DC programming approach to the minimum sum-of-squares clustering problems. *Pattern Recognit.* **2016**, *53*, 12–24. [CrossRef]
30. Karmitsa, N.; Bagirov, A.M.; Taheri, S. Clustering in large data sets with the limited memory bundle method. *Pattern Recognit.* **2018**, *83*, 245–259. [CrossRef]
31. Bagirov, A.; Taheri, S.; Asadi, S. A difference of convex optimization algorithm for piecewise linear regression. *J. Ind. Manag. Optim.* **2019**, *15*, 909–932. [CrossRef]
32. Sayah, S.; Zehar, K. Modified differential evolution algorithm for optimal power flow with non-smooth cost functions. *Energy Convers. Manag.* **2008**, *49*, 3036–3042. [CrossRef]
33. Al-Sultan, K.S. A Tabu search approach to the clustering problem. *Pattern Recognit.* **1995**, *28*, 1443–1451. [CrossRef]
34. Oonsivilai, A.; Srisuruk, W.; Marungsri, B.; Kulworawanichpong, T. Tabu Search Approach to Solve Routing Issues in Communication Networks. *Int. J. Electr. Comput. Energ. Electron. Commun. Eng.* **2009**, *3*, 1211–1214.
35. Hemamalini, S.; Simon, S.P. Artificial Bee Colony Algorithm for Economic Load Dispatch Problem with Non-smooth Cost Functions. *Electr. Power Components Syst.* **2010**, *38*, 786–803. [CrossRef]
36. Niknam, T.; Mojarrad, H.D.; Meymand, H.Z.; Firouzi, B.B. A new honey bee mating optimization algorithm for non-smooth economic dispatch. *Energy* **2011**, *36*, 896–908. [CrossRef]
37. Basu, M. Modified Particle Swarm Optimization for Non-smooth Non-convex Combined Heat and Power Economic Dispatch. *Electr. Power Components Syst.* **2015**, *43*, 2146–2155. [CrossRef]
38. Schlüter, M.; Egea, J.A.; Banga, J.R. Extended ant colony optimization for non-convex mixed integer nonlinear programming. *Comput. Oper. Res.* **2009**, *36*, 2217–2229. [CrossRef]
39. Corazza, M.; Fasano, G.; Gusso, R. Particle Swarm Optimization with non-smooth penalty reformulation, for a complex portfolio selection problem. *Appl. Math. Comput.* **2013**, *224*, 611–624. [CrossRef]
40. Clarke, G.; Wright, J. Scheduling of Vehicles from a Central Depot to a Number of Delivery Points. *Oper. Res.* **1964**, *12*, 568–581. [CrossRef]
41. Bellmore, M.; Nemhauser, G.L. The traveling salesman problem: A survey. *Oper. Res.* **1968**, *16*, 538–558. [CrossRef]
42. Panwalkar, S.S.; Iskander, W. A survey of scheduling rules. *Oper. Res.* **1977**, *25*, 45–61. [CrossRef]
43. Juan, A.A.; Faulin, J.; Ferrer, A.; Lourenço, H.R.; Barrios, B. MIRHA: multi-start biased randomization of heuristics with adaptive local search for solving non-smooth routing problems. *Top* **2013**, *21*, 109–132. [CrossRef]
44. Resende, M.G.; Ribeiro, C.C. Greedy randomized adaptive search procedures: Advances, hybridizations, and applications. In *Handbook of Metaheuristics*; Springer: Berlin/Heidelberg, Germany, 2010; pp. 283–319.
45. Estrada-Moreno, A.; Fikar, C.; Juan, A.A.; Hirsch, P. A biased-randomized algorithm for redistribution of perishable food inventories in supermarket chains. *Int. Trans. Oper. Res.* **2019**, *26*, 2077–2095. [CrossRef]
46. Ferone, D.; Hatami, S.; González-Neira, E.M.; Juan, A.A.; Festa, P. A biased-randomized iterated local search for the distributed assembly permutation flow-shop problem. *Int. Trans. Oper. Res.* **2019**. [CrossRef]
47. Mazza, D.; Pages-Bernaus, A.; Tarchi, D.; Juan, A.A.; Corazza, G.E. Supporting mobile cloud computing in smart cities via randomized algorithms. *IEEE Syst. J.* **2016**, *12*, 1598–1609. [CrossRef]
48. Melo, M.T.; Nickel, S.; Saldanha-Da-Gama, F. Facility location and supply chain management–A review. *Eur. J. Oper. Res.* **2009**, *196*, 401–412. [CrossRef]

49. Ahmadi-Javid, A.; Seyedi, P.; Syam, S.S. A survey of healthcare facility location. *Comput. Oper. Res.* **2017**, *79*, 223–263. [CrossRef]
50. De Armas, J.; Juan, A.A.; Marquès, J.M.; Pedroso, J.P. Solving the deterministic and stochastic uncapacitated facility location problem: from a heuristic to a simheuristic. *J. Oper. Res. Soc.* **2017**, *68*, 1161–1176. [CrossRef]
51. Correia, I.; Melo, T. Multi-period capacitated facility location under delayed demand satisfaction. *Eur. J. Oper. Res.* **2016**, *255*, 729–746. [CrossRef]
52. Estrada-Moreno, A.; Ferrer, A.; Juan, A.A.; Bagirov, A.; Panadero, J. A biased-randomised algorithm for the capacitated facility location problem with soft constraints. *J. Oper. Res. Soc.* **2019**, 1–17. [CrossRef]
53. Cordeau, J.F.; Laporte, G.; Savelsbergh, M.W.; Vigo, D. Vehicle routing. *Handbooks in Operations Research and Management Science*; Elsevier: Amsterdam, The Netherlands, 2007; Volume 14, pp. 367–428.
54. Adewumi, A.O.; Adeleke, O.J. A survey of recent advances in vehicle routing problems. *Int. J. Syst. Assur. Eng. Manag.* **2018**, *9*, 155–172. [CrossRef]
55. Corberan, A.; Prins, C. Recent results on Arc Routing Problems: An annotated bibliography. *Networks* **2010**, *56*, 50–69. [CrossRef]
56. Corberan, A.; Laporte, G. *Arc Routing: Problems, Methods, and Applications*; SIAM: Philadelphia, PA, USA, 2013.
57. De Armas, J.; Ferrer, A.; Juan, A.A.; Lalla-Ruiz, E. Modeling and solving the non-smooth arc routing problem with realistic soft constraints. *Expert Syst. Appl.* **2018**, *98*, 205–220. [CrossRef]
58. Gonzalez, S.; Juan, A.A.; Riera, D.; Castella, Q.; Munoz, R.; Perez, A. Development and assessment of the SHARP and RandSHARP algorithms for the arc routing problem. *AI Commun.* **2012**, *25*, 173–189. [CrossRef]
59. Pinedo, M.L. *Scheduling: Theory, Algorithms, and Systems*, 3rd ed.; Springer: Berlin/Heidelberg, Germany, 2008.
60. Ferrer, A.; Guimarans, D.; Ramalhinho, H.; Juan, A.A. A BRILS metaheuristic for non-smooth flow-shop problems with failure-risk costs. *Expert Syst. Appl.* **2016**, *44*, 177–186. [CrossRef]
61. Ferrer, A.; Bagirov, A.; Beliakov, G. Solving DC programs using the cutting angle method. *J. Glob. Optim.* **2015**, *61*, 71–89. [CrossRef]

© 2019 by the authors. Licensee MDPI, Basel, Switzerland. This article is an open access article distributed under the terms and conditions of the Creative Commons Attribution (CC BY) license (http://creativecommons.org/licenses/by/4.0/).

Article

Planning the Schedule for the Disposal of the Spent Nuclear Fuel with Interactive Multiobjective Optimization

Outi Montonen [1,*], Timo Ranta [1,2] and Marko M. Mäkelä [1]

[1] Department of Mathematics and Statistics, University of Turku, FI-20014 Turku, Finland; timo.ranta@utu.fi (T.R.); makela@utu.fi (M.M.M.)
[2] Laboratory of Mathematics, Tampere University of Technology, FI-28100 Pori, Finland
* Correspondence: outi.montonen@utu.fi

Received: 24 October 2019; Accepted: 22 November 2019; Published: 25 November 2019

Abstract: Several countries utilize nuclear power and face the problem of what to do with the spent nuclear fuel. One possibility, which is under the scope in this paper, is to dispose of the fuel assemblies in the disposal facility. Before the assemblies can be disposed of, they must cool down their decay heat power in the interim storage. Next, they are loaded into canisters in the encapsulation facility, and finally, the canisters are placed in the disposal facility. In this paper, we model this process as a nonsmooth multiobjective mixed-integer nonlinear optimization problem with the minimization of nine objectives: the maximum number of assemblies in the storage, maximum storage time, average storage time, total number of canisters, end time of the encapsulation, operation time of the encapsulation facility, the lengths of disposal and central tunnels, and total costs. As a result, we obtain the disposal schedule i.e., amount of canisters disposed of periodically. We introduce the interactive multiobjective optimization method using the two-slope parameterized achievement scalarizing functions which enables us to obtain systematically several different Pareto optimal solutions from the same preference information. Finally, a case study adapting the disposal in Finland is given. The results obtained are analyzed in terms of the objective values and disposal schedules.

Keywords: achievement scalarizing functions; interactive method; multiobjective optimization; nonsmooth optimization; spent nuclear fuel disposal

1. Introduction

The disposal of the spent nuclear fuel is a challenging task where the careful planning and optimization of processes definitely pays dividends. The difficulty of the decision making is increased also by the fact that the disposal continues for the distant future and many parameters are still unknown. Indeed, the decisions made now have long term consequences. Thus, it is only reasonable to investigate different scenarios by utilizing multiobjective optimization from the different perspectives.

The disposal is a topical issue since many of the countries utilizing nuclear power have not yet disposed of any spent nuclear fuel. Nevertheless, all of them have to do something for it sooner or later. Long-term storage in interim storage is not considered a safe or ethical solution [1]. At the same time, the geological disposal is stated to be widely accepted as a safe method [1]. Finland is going to be one of the first countries to dispose of the spent nuclear fuel by starting the disposal in 2020s [2].

The aim in the geological disposal is to isolate the spent nuclear fuel to the bedrock such that it has no more impacts on the environment than the regular background radiation. First, the fuel assemblies are removed from the reactor and stored in the water pool in the reactor hall in order to decrease the radiation and the decay heat power to the suitable level such that the assemblies can be transferred to the water pool in the interim storage facility for decades. When the assemblies are cool enough, they can be transferred to the encapsulation facility, where the assemblies are encapsulated into the copper-iron canisters. After that, the canister moves on towards the disposal facility, in depth of more than 400 m. The disposal facility consists of the central tunnel and several parallel disposal tunnels that are connected to the central tunnel. The canister is placed vertically in the hole on the floor of the disposal tunnel. Finally, the disposal tunnel is filled up and sealed. In this study, we divide the disposal process into three parts: the interim storage, the encapsulation facility, and the disposal facility.

As the entire nuclear waste management is a large task, optimization related studies about it are usually focused on some smaller entities. Some of these entities are concentrated more on political or social aspects like to determine where to put a disposal repository [3] or how to route the transfer of the nuclear waste, or hazardous waste in general [4,5]. More safety-related aspects are the optimization of the nuclear safeguards [6,7] and the safety assessment of nuclear waste repositories [8]. In our study, we aim to produce a disposal schedule such that several goals related to all the interim storage, the encapsulation facility, and the disposal facility are taken into account simultaneously with multiobjective optimization. Other studies aiming at a disposal schedule are, for example, [9] where a single-objective mixed-integer linear programming (MILP) model minimizing the costs is given and [10] trying to achieve the minimal area of the disposal facility with a linear transportation model. Another research related to the disposal facility is discussed in [11], where the multiobjective MILP problem is given to optimize the nuclear waste placement in the disposal facility. In addition, there are attempts to optimize the loading of canisters in Finland [9,12], Slovenia [13], and Switzerland [14].

This study continues the work of [9], where the aim was to minimize the total costs of the disposal in Finland by selecting the schedule of the disposal. Here, this work has been continued by remodeling the situation as the nonsmooth multiobjective mixed-integer nonlinear programming (MINLP) problem. As a nonsmooth optimization problem [15–17], the objectives and the constraints are not necessarily continuously differentiable functions. This allows us to model the situation more accurately. Indeed, many practical applications have nonsmooth nature (see e.g., [18–20]) even if they are modeled as differentiable problems in many cases in practice.

Many practical problems also involve several objectives [21–24]. As a problem of this scale, this application has several conflicting objectives to offer naturally. Besides total costs, it is reasonable to optimize, for instance, the area of the disposal facility. In our model, this is done by minimizing the lengths of both disposal and central tunnels. In total, our model contains nine objectives. In addition to the previous three objectives, we have three objectives related to the interim storage and three related to the encapsulation facility. In the interim storage, we want to minimize the maximum number of assemblies in the storage, the maximum storage time, and the average storage time. On the other hand, the operation time of the encapsulation facility is aimed to be minimized and al number of canisters, or in other words, the number of the empty assembly positions.

These objectives indeed are conflicting. For instance, we want the whole disposal process to be over as early as possible, but this raises the heat production load of the canister. This in its turn, increases the distances between the canisters in the disposal facility. However, the heat load of the canister can be decreased by leaving empty assembly positions, but then more canisters are needed. Another option is to increase the cooling time which again delays the end of the disposal, but if the disposal delays, more storage space is needed. Obviously, all of these decisions have an impact on costs. As exemplified, the minimization of only one objective may lead to an unsatisfactory solution with respect to some other objective. This leads us to a situation where compromises are certainly needed.

As a result of the multiobjective optimization, we obtain several mathematically equally good compromises, called Pareto optimal solutions. The final selection is left to the decision-maker who has more insight into the problem. In this paper, we propose an interactive procedure utilizing the achievement scalarizing function (ASF), in particular, the two-slope parameterized ASF [25] which bases on parameterized ASF [26] and two-slope ASF [27] generalizing both of them by combining their advantages. Via scalarization, the original multiobjective problem is transformed into one single-objective problem. The idea in brief with ASF is that the decision-maker gives a reference point including the decision maker's wishes towards the final solution. Then, the closest optimal solution with respect to some metric is found. If we use only one metric, as is the case in general with ASF, the selected metric defines which solution is found [28–30]. With the parameterization, we are able to use several metrics, nine in this particular case, and thus, yield different solutions with reasonable distribution. This ability to systematically generate different solutions from the same preference information is utilized in the interactive framework.

This paper is organized as follows. In Section 2, we begin by depicting the situation under the consideration and give a nonsmooth multiobjective MINLP model for it. In Section 3, we first introduce some fundamental preliminaries about multiobjective optimization, and then describe the multiobjective interactive method utilizing two-slope parameterized ASFs (MITSPA). In Section 4, one special case study of the disposal in Finland is given and the solutions are analyzed. Finally, in Section 5 some concluding remarks are discussed.

2. Mathematical Model

In this section, we give a comprehensive description of the model for scheduling the disposal of the spent nuclear fuel. The aim is to provide general guidelines for the disposal schedule and we only plan how many canisters are disposed of rather than which assemblies are placed in which canister nor give any complex lay-out for the disposal facility. We model the situation adapting the disposal in Finland as described in the introduction with some limitations like we omit the transportation between the facilities. Furthermore, we suppose that nothing is disposed of yet and only one type of fuel is considered. Some other simplifying assumptions are that we have access to all the assemblies, assemblies are identical, and the bedrock is homogeneous such that we can build tunnels anywhere.

The model formulated is a nonsmooth multiobjective MINLP problem having nine objectives. One obvious objective is total costs. Due to the long term time perspective of the disposal, the costs will probably change during the years so we minimize also some cost factors as their own objectives. Besides being a cost factor, these objectives have also other reasons to be selected as an objective. The interim storage-related objectives minimize storage times and amounts. The faster the assemblies get under the ground, the safer it is. Other safety issues are handled as constraints, like the cooling time of the assembly must be sufficient, the maximum decay heat power of the canister is limited, and the distances between disposal tunnels and canisters depend on the heat load of the canister. While we allow empty positions in canisters, we still try to keep the total amount of the canisters as low as possible. The other objectives related to the encapsulation facility aim to get disposal done as soon as possible. Finally, the area of the disposal facility is minimized.

2.1. Parameters

The model involves several parameters mostly dealing with lower and upper bounds and costs. First, we begin with two parameters determining the size of the model. Let

N be a total number of disposal periods
Z be a total number of removals from the reactor.

In addition, we define two sets of indices: the set of periods $\mathcal{N} = \{1, \ldots, N\}$ and the set of removals from the reactor $\mathcal{Z} = \{1, \ldots, Z\}$. Note that part of the removals are done before the first disposal period begins. In order to link the removals from the reactor and periods, we introduce two parameters:

a the last removal before the first disposal period
b the disposal period when the last removal is done.

In the following, we specify notation and measurement units for some physical magnitudes:

M_i number of assemblies belonging to the removal $i \in \mathcal{Z}$
Q length of one disposal tunnel [m]
$A_{i,j}$ storage time of an assembly belonging to the removal $i \in \mathcal{Z}$
      in the period $j \in \mathcal{N}$ [period]
$P_{i,j}$ decay heat power of an assembly belonging to the removal $i \in \mathcal{Z}$
      in the period $j \in \mathcal{N}$ [W].

The next seven parameters describe the cost information needed as an input data for the model:

C_{AS} storage cost per one assembly per period [€]
C_{IS} costs related to the interim storage per period [€]
C_{SP} cost of a storage place per one assembly [€]
C_{CA} cost of one canister [€]
C_{EF} costs related to operating the encapsulation facility per period [€]
C_{DT} cost of a disposal tunnel per meter [€]
C_{CT} cost of a central tunnel per meter [€].

Finally, we give some parameters related to the upper and lower bounds:

R minimum storage time of an assembly [period]
K maximum capacity of a canister
T minimum number of canisters disposed in one period
U maximum number of canisters disposed in one period
$p_{max}^{low}, p_{max}^{up}$ lower and upper bound for the maximum average power of
      a canister [W]
$d_{CA}^{low}, d_{CA}^{up}$ lower and upper bound for the distance between canisters [m]
$d_{DT}^{low}, d_{DT}^{up}$ lower and upper bound for the distance between disposal
      tunnels [m].

2.2. Continuous Variables

The model involves $N(2Z+1)+3$ continuous variables such that they all are assumed to be non-negative. The continuous variables used are:

$x_{i,j}$ number of assemblies belonging to the removal $i \in \mathcal{Z}$ disposed during
      the period $j \in \mathcal{N}$
y_j number of canisters disposed during the period $j \in \mathcal{N}$
$z_{i,j}$ number of assemblies belonging to the removal $i \in \mathcal{Z}$ being in storage
      at the end of the period $j \in \mathcal{N}$
p_{max} maximum average power of a canister
d_{DT} distance between two adjacent disposal tunnels
d_{CA} distance between two adjacent canisters in a disposal tunnel.

Note that the first three variables have integer nature, but in order to ease the computation, they are relaxed as continuous variables.

2.3. Binary Variables

Besides continuous variables, the model consists also $N(2Z+3)$ binary variables listed below:

e_{ON}^j encapsulation starts in the beginning of the period $j \in \mathcal{N}$
e_{OFF}^j encapsulation ends in the beginning of the period $j \in \mathcal{N}$
e_j encapsulation facility is in operation during the period $j \in \mathcal{N}$
$s_{i,j}$ assemblies belonging to the removal $i \in \mathcal{Z}$ take off from disposal
      at the beginning of the period $j \in \mathcal{N}$
$r_{i,j}$ indicates that assemblies belonging to the removal $i \in \mathcal{Z}$
      can be disposed during the period $j \in \mathcal{N}$.

2.4. Objectives

The model involves nine objectives such that six of them are nonlinear and three are linear. These objectives are:

$$\min \quad \max\left\{ \sum_{i=1}^{a} M_i, \sum_{i=1}^{a+j} z_{i,j}, \sum_{i \in \mathcal{Z}} z_{i,l} \,\Big|\, j \in \{1,\ldots,b-1\}, l \in \{b,\ldots,N\} \right\} \quad (1)$$

$$\min \quad \max\{ A_{i,j} s_{i,j} - 1 \mid i \in \mathcal{Z}, j \in \mathcal{N} \} \quad (2)$$

$$\min \quad \frac{\sum_{i \in \mathcal{Z}} \sum_{j \in \mathcal{N}} A_{i,j} x_{i,j}}{\sum_{i \in \mathcal{Z}} M_i} \quad (3)$$

$$\min \quad \sum_{j \in \mathcal{N}} y_j \quad (4)$$

$$\min \quad \max\{ e^j_{OFF} \cdot j - 1 \mid j \in \mathcal{N} \} \quad (5)$$

$$\min \quad \sum_{j \in \mathcal{N}} e_j \quad (6)$$

$$\min \quad d_{CA} \sum_{j \in \mathcal{N}} y_j \quad (7)$$

$$\min \quad \frac{1}{Q} d_{CA} d_{DT} \sum_{j \in \mathcal{N}} y_j \quad (8)$$

$$\begin{aligned}
\min \quad & C_{AS} \sum_{i \in \mathcal{Z}} \sum_{j \in \mathcal{N}} A_{i,j} x_{i,j} + C_{IS} \max\{ e^j_{OFF} \cdot j - 1 \mid j \in \mathcal{N} \} \\
+ \quad & C_{SP} \max\left\{ \sum_{i=1}^{a} M_i, \sum_{i=1}^{a+j} z_{i,j}, \sum_{i \in \mathcal{Z}} z_{i,l} \,\Big|\, j \in \{1,\ldots,b-1\}, l \in \{b,\ldots,N\} \right\} \\
+ \quad & C_{CA} \sum_{j \in \mathcal{N}} y_j + C_{EF} \sum_{j \in \mathcal{N}} e_j + C_{DT} d_{CA} \sum_{j \in \mathcal{N}} y_j + C_{CT} \tfrac{1}{Q} d_{DT} d_{CA} \sum_{j \in \mathcal{N}} y_j.
\end{aligned} \quad (9)$$

Note that from nonlinear objectives, the objectives (1), (2), (5) and (9) are also nonsmooth. The objectives (1)–(3) are related to the interim storage such that (1) minimizes the maximum number of assemblies in the storage, (2) minimizes the maximum storage time, and (3) minimizes the average storage time. In the objective (1), with the first component we take into account the first a removals from the reactor where all the assemblies must be stored simultaneously. The second component handles the cases when removals are accomplished during the disposal periods. Finally, with the third component the cases when all removals are done are considered.

The next three objectives (4)–(6) are related to the encapsulation facility. The objective (4) minimizes the total number of canisters, (5) aims to stop the disposal as early as possible, and (6) minimizes the time which the encapsulation facility is in operation.

The objectives (7) and (8) aim to minimize the size of the disposal facility such that (7) minimizes the total length of disposal tunnels and (8) minimizes the length of the central tunnel.

Finally, the ninth objective (9) minimizes the total costs of the disposal process. The costs taken into account are related to the storage, cost of individual canisters, the encapsulation facility operating costs, and the building costs of the disposal and central tunnels.

2.5. Constraints—Interim Storage

The first set of constraints are related to the interim storage. All of these $Z(5N+2) + N + 2$ constraints are linear.

$$z_{i,1} - M_i + x_{i,1} = 0, \quad i \in \mathcal{Z} \tag{10}$$

$$z_{i,j} - z_{i,j-1} + x_{i,j} = 0, \quad i \in \mathcal{Z}, j \in \mathcal{N} \setminus \{1\} \tag{11}$$

$$z_{i,N} = 0, \quad i \in \mathcal{Z} \tag{12}$$

$$\sum_{j \in \mathcal{N}} s_{i,j} = 1 \quad i \in \mathcal{Z} \tag{13}$$

$$r_{i,1} = e^1_{ON} - s_{i,1}, \quad i \in \mathcal{Z} \tag{14}$$

$$r_{i,j} = r_{i,j-1} + e^j_{ON} - s_{i,j}, \quad i \in \mathcal{Z}, j \in \mathcal{N} \setminus \{1\} \tag{15}$$

$$r_{i,j} \leq e_j, \quad i \in \mathcal{Z}, j \in \mathcal{N} \tag{16}$$

$$x_{i,j} \leq U K r_{i,j}, \quad i \in \mathcal{Z}, j \in \mathcal{N} \tag{17}$$

$$x_{i,j}(A_{i,j} - R) \geq 0, \quad i \in \mathcal{Z}, j \in \mathcal{N} \tag{18}$$

The constraints (10)–(12) define the variables $z_{i,j}$ depicting the amount of assemblies in storage. The constraint (13) enforces all the assemblies to be disposed once. With the constraints (14)–(16) the variables $r_{i,j}$ are defined. The constraint (17) ensures that the production capacity is not exceeded and the constraint (18) ensures that the assembly disposed has been cooling long enough.

2.6. Constraints—Encapsulation Facility

In order to guarantee the acceptable encapsulation, the following $4N + 1$ linear constraints are needed.

$$\sum_{j \in \mathcal{N}} e^j_{ON} = 1 \tag{19}$$

$$\sum_{j \in \mathcal{N}} e^j_{OFF} = 1 \tag{20}$$

$$e_1 = e^1_{ON} - e^1_{OFF} \tag{21}$$

$$e_j = e_{j-1} + e^j_{ON} - e^j_{OFF}, \quad j \in \mathcal{N} \setminus \{1\} \tag{22}$$

$$y_j \geq \frac{1}{K} \sum_{i \in \mathcal{Z}} x_{i,j}, \quad j \in \mathcal{N} \tag{23}$$

$$y_j \leq U e_j, \quad j \in \mathcal{N} \tag{24}$$

$$y_j \geq T(e_j - e^{j+1}_{OFF}), \quad j \in \mathcal{N} \setminus \{N\}. \tag{25}$$

The constraints (19) and (20) ensures that the encapsulation facility is switched on and off exactly once meaning that all the canisters must be encapsulated at once. The constraints (21) and (22) define the variable e_j. The constraints (23)–(25) guide the encapsulation process: (23) guarantees that there exist enough canisters such that all the assemblies can be disposed, (24) keeps the number of canisters under the production capacity, and (25) forces the minimum production to be fulfilled.

2.7. Constraints—Disposal Facility

The number of constraints related to the disposal facility is $N+4$ such that $N+1$ of them are nonlinear, and three of the constraints are box constraints.

$$\sum_{i \in Z} P_{i,j} x_{i,j} - p_{max} y_j \leq 0, \quad j \in \mathcal{N} \tag{26}$$

$$d_{CA} - g(p_{max}, d_{DT}) = 0 \tag{27}$$

$$p_{max} \in \left[p_{max}^{low}, p_{max}^{up} \right] \tag{28}$$

$$d_{CA} \in \left[d_{CA}^{low}, d_{CA}^{up} \right] \tag{29}$$

$$d_{DT} \in \left[d_{DT}^{low}, d_{DT}^{up} \right] \tag{30}$$

The constraints (26) and (27) are the nonlinear constraints of this model. The constraint (26) ensures that the heat power of the canisters disposed is allowable while the constraint (27) defines the dependence between the variables d_{CA}, p_{max}, and d_{DT}. In our case, this nonlinear function $g : \mathbb{R}^2 \to \mathbb{R}$ is approximated with a piece-wise linear function (see Appendix A). Finally, the box constraints (28)–(30) give lower and upper bounds for variables p_{max}, d_{CA}, and d_{DT}, respectively.

Finally, we give some boundaries for the variables:

$$x_{i,j} \geq 0, \quad z_{i,j} \geq 0 \quad \text{for all} \quad i \in Z, j \in \mathcal{N},$$
$$y_j \geq 0 \quad \text{for all} \quad j \in \mathcal{N},$$
$$e_{ON}^j \in \{0,1\}, \quad e_{OFF}^j \in \{0,1\}, \quad e_j \in \{0,1\} \quad \text{for all} \quad j \in \mathcal{N}$$
$$s_{i,j} \in \{0,1\}, \quad r_{i,j} \in \{0,1\} \quad \text{for all} \quad i \in Z, j \in \mathcal{N}.$$

To conclude, the model has nine objectives such that 6 are nonlinear and 3 are linear. The rest of the dimensions of the model are depending on two parameters: the number of periods N and the number of the removals from the reactor Z. Number of constraints is $5(N(Z+1)+1)+2Z$, where are $Z(5N+2)+4N+1$ linear, $N+1$ nonlinear and 3 box constraints. The total number of variables is $4N(Z+1)+3$ and $N(2Z+1)+3$ of them are non-negative continuous variables and $N(2Z+3)$ are binary variables. Evidently, with any realistic values of N and Z, for example $N=19$ and $Z=11$ when one period is five years, the size of the problem will come quite large.

3. Multiobjective Optimization Approach

In this section, we define some fundamental aspects on multiobjective optimization, and then, describe the family of two-slope parameterized achievement scalarizing functions (ASFs) [25] with its properties. Finally, the interactive method utilizing two-slope parameterized ASFs is introduced.

3.1. Mathematical Background

We consider the following multiobjective MINLP problem of the form

$$\min_{x \in X} f(x) = \{f_1(x), \ldots, f_k(x)\}, \tag{31}$$

where $x \in X = \{x = (y,z) \mid y \in \mathbb{R}^n, z \in \mathbb{Z}^m\} \cap C$ is a decision variable, C is the set of constraints, and X is a nonempty and compact set of feasible solutions. The objectives $f_i : X \to \mathbb{R}$ for all $i \in I = \{1, \ldots, k\}$ are assumed to be lower semicontinuous with respect to y and at least partially conflicting. Therefore, we cannot find a minimal solution for every objective simultaneously and the minimization of only one objective may lead to an arbitrary bad solution with respect to other objectives. In order to compare the objectives, for $x, y \in \mathbb{R}^k$ we denote by $x < y$ if $x_i < y_i$ for all $i \in I$ and $x \leq y$ if $x_i \leq y_i$ for all $i \in I$.

In multiobjective optimization, we say that a solution is Pareto optimal if we cannot improve any objective without causing a deterioration for some other objective at the same time. Mathematically speaking, a solution $x^* \in X$ is Pareto optimal if there does not exist any solution $x \in X$ such that $f(x) \leq f(x^*)$ and $f_j(x) < f_j(x^*)$ for at least one index $j \in I$. It is noteworthy that usually we do not have a unique Pareto optimum but a set of Pareto optimal solutions, called the Pareto set. All these Pareto optimal solutions belong also to a larger class of weakly Pareto optimal solutions. The solution $x' \in X$ is an element of this class if there does not exist another solution $x \in X$ such that $f(x) < f(x')$.

In order to obtain some information about the Pareto set, we can define an ideal and a nadir vector, $f^{id} \in \mathbb{R}^k$ and $f^{nad} \in \mathbb{R}^k$, to give the lower and the upper bound for a Pareto optimal solution, respectively. The ideal vector consists of individual minima of the objectives. This means that the component f_i^{id} is calculated as a solution of the problem $\min_{x \in X} f_i(x)$. Due to the conflicting objectives, the ideal vector is not feasible. The nadir vector, in its turn, represents the worst objective values in the Pareto set. Unfortunately, the exact calculation of the nadir vector needs the maximization of objectives over the set of Pareto optimal solutions being a hard task. Thus, the nadir vector needs to be approximated, for example, with a pay-off table (see e.g., [31,32]).

3.2. Two-Slope Parameterized ASFs

We approach the multiobjective mixed-integer problem with a special type of achievement scalarizing functions. In general, the utilization of the achievement scalarizing function (ASF) aims to find a Pareto optimal solution being as close as possible to a so-called reference point f^R. The components $f_i^R, i \in I$ include the decision maker's wishes for each objective. This search is done by transforming the multiobjective optimization problem to a certain type of a scalarized problem and then applying some suitable single-objective optimization method.

We use here the two-slope parameterized ASF, proposed in [25], which is a generalization of the parameterized ASF [26] and the two-slope ASF [27]. Usually, to find the closest point to the reference point f^R, the distance from f^R is measured with only one metric. With the parameterization used in the parameterized ASF and the two-slope parameterized ASF, we can combine different metrics such that L_∞ and L_1 metrics are the extreme cases. Thus, by systematically producing different Pareto optimal solutions from the same preference information, we can give the decision maker a wider perspective to the range of Pareto optimal solutions. Another benefit of the two-slope parameterized ASF, as well as the two-slope ASF, is that we do not need to test the achievability of the reference point. This is due to the fact that the different weights are used depending on if the reference point is achievable (i.e., the reference point belongs to the image of the feasible solutions in the objective space) or unachievable. The use of different weights is reasonable since the decision-maker usually prefers different solutions if the reference point is achievable or not, as was suggested in [28].

In order to solve the model described in Section 2, we apply the two-slope parameterized ASF. Once the multiobjective problem is converted to the single-objective one, we obtain a scalarized version of the problem (31) in the form [25]

$$\min_{x \in X} \max_{\substack{I^q \subseteq I \\ |I^q|=q}} \left\{ \sum_{i \in I^q} \left[\max\{\lambda_i^U (f_i(x) - f_i^R), 0\} + \min\{\lambda_i^A (f_i(x) - f_i^R), 0\} \right] \right\}, \quad (32)$$

where the weighting vectors $\lambda_i^U, \lambda_i^A > 0$ for all $i \in I$ are for the unachievable and the achievable reference point, respectively. The parameter $q \in I$ specifies which metric is used and I^q is a set containing q integers from the interval $[i,k]$, where k is the total number of objectives. Then, the maximization is taken over all the sets including q integers from the interval $[1,k]$. In order to gain the benefits of the parameterization, or in other words, to use more metrics than only L_1 (i.e., $q = k$) and L_∞ (i.e., $q = 1$), the problem must contain at least three objectives while the maximum number of different metrics equals the number of the objectives.

Next, we are interested to know what can be deduced from the optimal solution of the scalarized problem. As the justification for the use of the two-slope parameterized ASF, we can proof the following results by adapting the proofs from [25].

Theorem 1 ([25]). *For the scalarized problem* (32) *it holds that:*

(i) *Any optimal solution of the scalarized problem is weakly Pareto optimal for the problem* (31).
(ii) *Among optimal solutions of the scalarized problem, there exists at least one Pareto optimal solution for the problem* (31).
(iii) *If x^* is a weakly Pareto optimal solution for the problem* (31), *then it is a solution of the scalarized problem* (32) *with $f^R = f(x^*)$, and the optimal value is zero.*

Proof. (i) Assume that x^* is an optimal solution of the problem (32) but not a weakly Pareto optimal solution of the problem (31). Then there exists a feasible solution $x' \in X$ such that $f(x') < f(x^*)$. For any $x \in X$, denote $I_x = \{i \in I^q \mid f_i^R \leq f_i(x)\}$, $J_x = \{i \in I^q \mid f_i^R > f_i(x)\}$ and $s_R^q(f(x), \lambda^U, \lambda^A)$ as the objective of the scalarized problem (32). Now

$$s_R^q(f(x'), \lambda^U, \lambda^A) = \max_{\substack{I^q \subseteq I \\ |I^q| = q}} \left\{ \sum_{i \in I_{x'}} \lambda_i^U(f_i(x') - f_i^R) + \sum_{i \in J_{x'}} \lambda_i^A(f_i(x') - f_i^R) \right\}$$

$$< \max_{\substack{I^q \subseteq I \\ |I^q| = q}} \left\{ \sum_{i \in I_{x'}} \lambda_i^U(f_i(x^*) - f_i^R) + \sum_{i \in J_{x'}} \lambda_i^A(f_i(x^*) - f_i^R) \right\}$$

$$\leq \max_{\substack{I^q \subseteq I \\ |I^q| = q}} \left\{ \sum_{i \in I_{x^*}} \lambda_i^U(f_i(x^*) - f_i^R) + \sum_{i \in J_{x^*}} \lambda_i^A(f_i(x^*) - f_i^R) \right\}$$

$$= s_R^q(f(x^*), \lambda^U, \lambda^A)$$

yielding to a contradiction.

(ii) Assume that x^* is an optimal solution of the problem (32) but not a Pareto optimal solution of the problem (31). Therefore, there exists $x' \in X$ such that $f(x') \leq f(x^*)$ and at least one index $j \in I$ such that $f_j(x') < f_j(x^*)$. Similarly to (i), we can deduce that $s_R^q(f(x'), \lambda^U, \lambda^A) \leq s_R^q(f(x^*), \lambda^U, \lambda^A)$. If the equality holds, x' is an optimal solution for the problem (32) and Pareto optimal for the problem (31). In the case of strict inequality, this yields to a contradiction with an assumption that x^* is an optimal solution for the problem (32).

(iii) First, we observe that s_R^q is strictly increasing (i.e., $s_R^q(f(x_1), \lambda^U, \lambda^A) < s_R^q(f(x_2), \lambda^U, \lambda^A)$ for any $f(x_1), f(x_2)$ having $f(x_1) < f(x_2)$ and $x_1, x_2 \in X$). Indeed, by taking $x_1, x_2 \in X$ such that $f(x_1) < f(x_2)$, we see that

$$s_R^q(f(x_1), \lambda^U, \lambda^A) = \max_{\substack{I^q \subseteq I \\ |I^q| = q}} \left\{ \sum_{i \in I_{x_1}} \lambda_i^U(f_i(x_1) - f_i^R) + \sum_{i \in J_{x_1}} \lambda_i^A(f_i(x_1) - f_i^R) \right\}$$

$$< \max_{\substack{I^q \subseteq I \\ |I^q| = q}} \left\{ \sum_{i \in I_{x_2}} \lambda_i^U(f_i(x_2) - f_i^R) + \sum_{i \in J_{x_2}} \lambda_i^A(f_i(x_2) - f_i^R) \right\}$$

$$= s_R^q(f(x_2), \lambda^U, \lambda^A).$$

The claim is obtained, since for any strictly increasing ASF it holds that a weakly Pareto optimal solution x^* for the problem (31) is a solution of the scalarized problem with $f^R = f(x^*)$ and the optimal value of s_R^q is zero (see [32]). □

Thus, we know that every Pareto optimal solution can be obtained and the solution of the scalarized problem (32) is weakly Pareto optimal for the original multiobjective problem. In order to guarantee the Pareto optimality of solutions, a so-called augmentation term [32]

$$\rho \sum_{i \in I} \lambda_i (f_i(x) - f_i^R), \quad \rho > 0 \tag{33}$$

may be added to the objective of the scalarized problem (32) [25]. Note that similarly to Theorem 8 in [25] it can be proven that if the set X and the objectives f_i, $i \in I$ are convex, then $s_R^q(f(x), \lambda^U, \lambda^A)$ preserves the convexity.

3.3. Multiobjective Interactive Method Utilizing the Two-Slope Parameterized ASFs

In the following, we state an outline of the multiobjective interactive solution approach utilizing the two-slope parameterized ASFs (MITSPA) applying reference point based preference information. The general framework of interactive methods is usually similar: firstly, some range for Pareto optimal solutions is given the decision-maker, secondly, the decision-maker provides some preference information, thirdly, some solutions are presented for the decision-maker, and fourthly, the decision-maker express his/her opinion on the solutions and modify the preference information as a base for the new solutions. The process is stopped when the decision-maker is satisfied with the solution. The main differences in various interactive methods can be found in the ways the preference information is given and which solvers are applied (see e.g., [32,33]).

Similar approaches to ours in terms of the utilization of scalarization functions and the reference point as preference information are proposed, for instance, in [32–38]. Compared with these, in our case with the two-slope parameterized ASFs, we can systematically produce different Pareto optimal solutions to obtain a reasonably distributed selection of Pareto set.

Multiobjective interactive method utilizing the two-slope parameterized ASFs (MITSPA)

Step 0. Give the ideal vector f^{id}, the nadir vector f^{nad}, and/or some Pareto optimal solution f_0 to the decision maker in order to illustrate the Pareto set.
Step 1. Set the iteration counter $h = 1$ and select the maximum number of iterations $h_{\max}$. Ask the decision maker to provide the reference point f_h^R and the number of solutions $s \in \{1, \ldots, k\}$ presented for each reference point. Initialize the positive coefficients λ^U and λ^A.
Step 2. Update the coefficients λ^U and λ^A if needed. Solve the problem (32) with the augmentation term (33) with the current reference point f_h^R.
Step 3. Present s solutions to the decision maker and ask the decision maker to select the most preferable solution among them as the current solution f_h and go to Step 5 or if more solutions for the current reference point f_h^R are needed go to Step 4.
Step 4. Present supplementary solutions to the decision maker. Ask the decision maker to select the most preferable solution among the previous s solutions and the supplementary solutions as the current solution f_h and go to Step 5.
Step 5. If $h = h_{\max}$ or the decision maker is satisfied with the current solution f_h, stop with the current solution as the final solution f^*. Otherwise, ask the decision maker to specify the new reference point f_{h+1}^R as the current reference point, set $h = h + 1$, and go to Step 2.

Some remarks about the above algorithm are in order. Step 0 consists of the illustration of the Pareto set. Some Pareto optimal solutions for the decision-maker to start with can be calculated, for example, by using the two-slope parameterized ASF (32) with an ideal vector as a reference point or by applying some suitable no-preference method like descent methods [39–43]. In Step 3, $s \in [1, k]$ solutions are presented to the decision-maker, where the k is the number of objectives. As mentioned, with the two-slope parameterized ASF we are able to solve as many different solutions as there are objectives. If the number of objectives is high, it facilitates the task of the decision-maker if only some of these solutions are presented. However, if the decision-maker is willing to see more solutions from the same reference point, this is enabled in step 4. If more than k solutions are needed in total for one reference point, they can be obtained by varying the coefficients λ^U and λ^A. During the solution process, the decision maker is able to learn about the model and after seeing some solutions, the decision maker has more insight into the problem and might want to change the opinion on the good reference point. Thus, in Step 5, a new reference point is allowed and new solutions are solved in Step 2.

4. Case Study: The Disposal in Finland

In practice, the scalarized problem (32) with an augmentation term (33) in Step 2 of MITSPA is solved with a branch-and-cut type method for single-objective MINLP problems called BARON [44,45] in GAMS [46]. The CPU time of solving each problem (32) presented here varies from 9 s to 28000 s while the average CPU time is 3475 s and the median CPU time is 142 s. The weighting vectors used are of the form

$$\lambda^U = \frac{1}{f^{nad} - f^R}, \quad \lambda^A = \frac{1}{f^R - f^{id}}$$

such that $f^{nad} - f^R > 0$ and $f^R - f^{id} > 0$ as suggested in [27]. The approximation of the nadir vector used is obtained with a pay-off table [31,32].

We investigate the disposal of the spent nuclear fuel from the European pressurized water reactor (EPR) produced by Olkiluoto 3 in Finland starting to operate in the near future. The length of one disposal period is selected to be 5 years, and the parameters N and Z are 19 and 11, respectively. The other parameters used are given in Appendix A, except the cost parameters that are omitted due to their commerce-related nature. This parameter selection yields a multiobjective MINLP problem with 9 objectives, 440 continuous and 475 binary variables, 1144 linear constraints, 20 nonlinear constraints, and three box constraints. Apart from these, we need some auxiliary variables and constraints to overcome the non-smoothness of the problem. Indeed, the two-slope parameterized ASFs are nonsmooth, but due to their min-max structure, the problem (32) can be written in the MINLP form as in [25]. Similarly, this trick can be applied also for the nonsmooth objectives. After that, we have to solve a single-objective problem with 441 continuous and 484 binary variables, 1153 linear constraints, 21–146 nonlinear constraints and 3 box constraints.

Before we proceed to the solution process, we discuss the trade-offs of the problem. There are three parts in the final disposal of spent nuclear fuel: the interim storage, the encapsulation facility, and the disposal facility. These three parts interact with each other as is exemplified in the following.

- Interim storage versus disposal facility: The interim storage-related goals all imply transferring the spent nuclear fuel from the interim storage as rapidly as possible. However, in order to minimize the disposal facility-related goals, the cooling times should be maximized.
- Encapsulation facility versus interim storage: By delaying the start of disposal, it is possible to shorten the operation time of the encapsulation facility, and thus, decrease the operating costs. Again, the delay at the start of the encapsulation can cause an increase of the inventories in the interim storage.
- Encapsulation facility versus disposal facility: The disposal should be started and ended as soon as possible. Both of these aims have a tendency to increase the canister heat load, and hence, affect the disposal facility goals. To minimize the operation time of the encapsulation facility, empty assembly positions can be used. However, the price to pay is the increased number of canisters. In addition, a larger number of canisters necessitates an increase in the disposal facility area.

In order to investigate these, and other trade-offs, the interactive method MITSPA is employed. In each iteration of MITSPA, some new preference information is asked from the decision-maker reflecting his/her preferences. For each iteration, we compute nine solutions by using the current reference point with different metrics by varying the value of the parameter q from 1 to 9 in Step 2. In order to exemplify this, the nine solutions computed using reference point 1 are shown in Figure 1. These nine solutions represent different trade-offs between objective function values. The results obtained are scaled to the interval from 0 to 1 such that 0 is the value of the ideal vector and 1 is the value of the nadir vector for the objective under consideration. The different solutions are labeled based on the reference point used and the value of the parameter q. For example, the solution r1q1 is the result obtained by using the reference point 1 and $q = 1$. Moreover, the reference point 1 is labeled with r1.

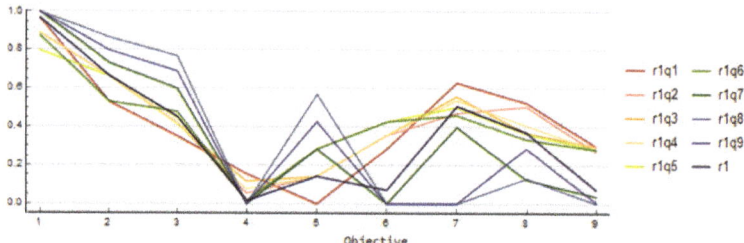

Figure 1. The objective values of 9 solutions obtained using the reference point 1.

In Step 3, two solutions are selected to be presented for the decision-maker for the closer inspection. The number of presented solutions s is restricted to two in order to aid the decision maker's task to select best out of only two options and in order to keep the presentation clear. At each iteration, one solution with a smaller value of q and one with a larger value of q are presented and different values of q are demonstrated in order to exemplify the variety of solutions. Next, we present four iterations of MITSPA.

Iteration 1. At the first iteration, the decision maker begins by investigating the trade-off between operation time of the encapsulation facility and the cooling times of assemblies by deciding to start with the unachievable reference point such that the operation time is short and the cooling time is long. The two solutions chosen for reference point 1 are shown in Figure 2a together with reference point 1. The solution obtained by using value $q = 1$ (r1q1) shown in the green line corresponds to the early starting time of disposal. The solution obtained by using value $q = 9$ (r1q9) shown in the orange line corresponds to the late starting time of the disposal. In Figures 2b,c, the corresponding disposal schedules are given. The solution r1q9, has the shortest possible encapsulation time but the maximum cooling time is long. The solution r1q9, like r1q1, has a high maximum number of assemblies in the storage (see the objective (1)), but the maximum and average storage times (the objectives (2) and (3)) are slightly shorter. The solution r1q9 does not allow any empty positions in canisters while the solution r1q1 does (see (4)), but the encapsulation ends much later (see (5)) in the solution r1q9 than in r1q1. However, the operation time of the encapsulation facility (see (6)) is shorter in the solution r1q9 than in r1q1. When the disposal facility-related objectives (see (7) and (8)) are compared, the solution r1q9 needs a smaller area than the solution r1q1. Moreover, the solution r1q9 is cheaper than the reference, while the solution r1q1 is more expensive than the reference (see (9)). Mainly due to the significant difference in the costs, the decision maker selects the solution r1q9 as the current solution f_1.

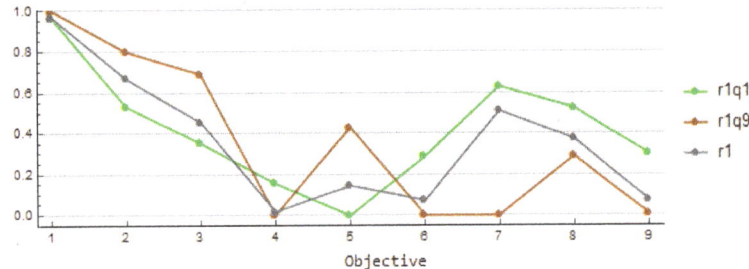

(a) The objective values for the selected solutions $q = 1$ and $q = 9$ and the reference point 1.

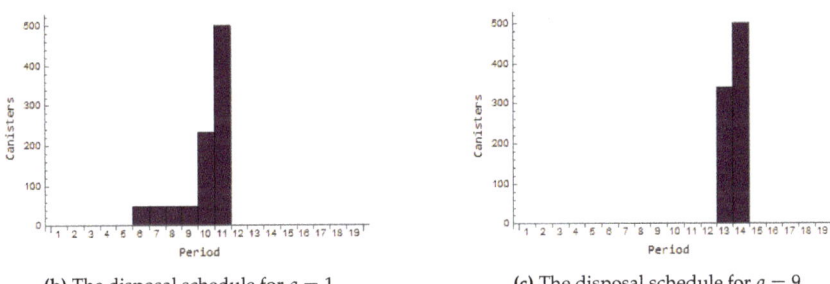

(b) The disposal schedule for $q = 1$.   (c) The disposal schedule for $q = 9$.

Figure 2. Results for the iteration 1.

Iteration 2. In order to learn more about the trade-off between the operation time of the encapsulation facility and the cooling time, another reference point (reference point 2) is selected. In this case, the reference point is achievable. Now we try to find solutions such that the operation time is longer and cooling time shorter. The two solutions chosen for reference point 2 are shown in Figure 3a and the corresponding disposal schedules in Figure 3b,c. Again, the solution obtained with the small value $q = 1$ (r2q1) represents the early starting time of the disposal. This is depicted with the green line in Figure 3a while the orange line depicts the solution obtained using value $q = 9$ (r2q9) corresponding to the late starting time of the disposal.

If we compare the disposal schedules in Figure 3b,c to the schedules for reference point 1 given in Figure 2b,c, we notice some similarity. Even though the starting and ending times differ as well as the total number of canisters, the solutions with the parameter q value 1 and value 9 have the same shape. The smaller q suggests the schedule such that first, we encapsulate a small number of canisters per period and the number of canisters is growing while the time goes by, whereas the larger q recommends the schedule where all the canisters are encapsulated within two periods. The solution r2q1 captures the reference point well since they coincide with respect to other objectives than the objectives (1) and (5) which are better than the reference values. Thus, the decision maker is willing to continue with the solution r2q1 as the current solution f_2.

Iteration 3. The long operation time of the encapsulation facility (the objective (6)) is still under the microscope at the third iteration but the decision-maker is tempted by the short central tunnel appeared in the previous iteration and combines the long operation time with small disposal facility area. Like the first reference point, also this is unachievable. The solution obtained by using $q = 2$ (r3q2) is shown in green and the solution obtained with $q = 8$ (r3q8) is shown in orange in Figure 4a. Figure 4b illustrates that the solution r3q2 yields a schedule with an early starting date and the disposal takes the longest time while the solution r3q8 starts the disposal later but it is performed faster as seen in Figure 4c. The solution r3q8 yields almost ideal value for the costs, and we can deduce that in order to achieve lower costs we have to give up in the objectives related to the storage capacity and times. Moreover, the disposal ends rather late. For the current solution f_3 the decision-maker selects the solution r3q8 due to the low costs and small disposal facility area.

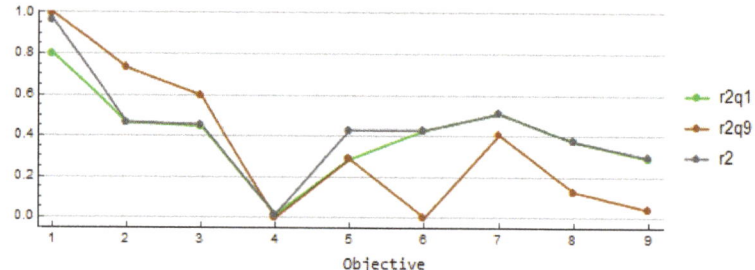

(a) The objective values for the selected solutions $q = 1$ and $q = 9$ and reference point 2.

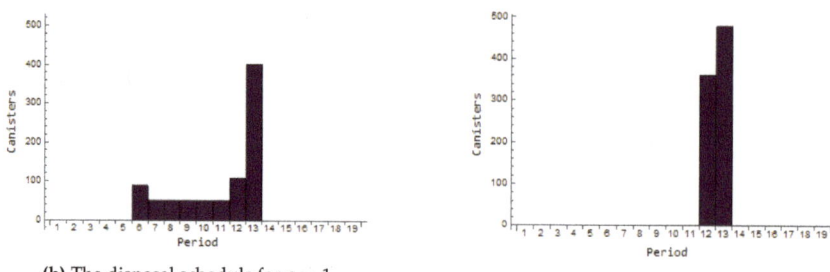

(b) The disposal schedule for $q = 1$.

(c) The disposal schedule for $q = 9$.

Figure 3. Results for the iteration 2.

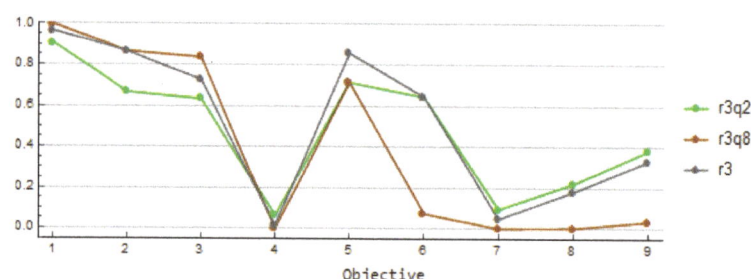

(a) The objective values for the selected solutions $q = 2$ and $q = 8$ and reference point 3.

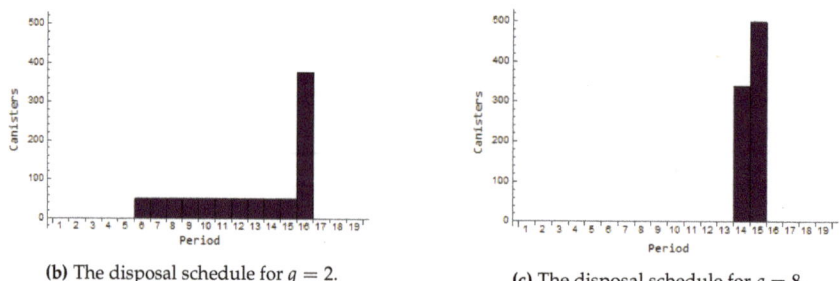

(b) The disposal schedule for $q = 2$.

(c) The disposal schedule for $q = 8$.

Figure 4. Results for the iteration 3.

Since one motivation for this research was to take into account more goals than just the costs, we are eager to see what happens if we omit the costs and solve the problem with only the first eight objectives (1)–(8). The reference point 3' is similar to the reference point 3 without the value for the costs. The results with $q = 2$ (r3'q2) and $q = 8$ (r3'q8) are given in Figure 5. Note that since there are now only eight objectives, the scalarized function is different than in the case of nine objectives. The solutions in Figure 5 are quite similar and there is less variation than in the solutions in Figure 4a.

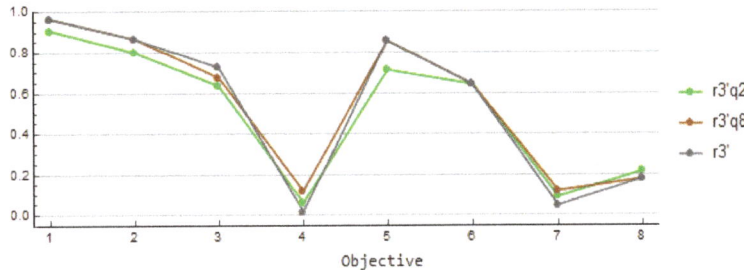

Figure 5. The objective values corresponding the selected solutions $q = 2$ and $q = 8$ for the modified reference point 3 with objectives (1)–(8).

Iteration 4. The current solution f_3 has high interim storage capacity and a small amount of canisters. At the fourth iteration, the decision maker is interested in to see if the opposite is possible, namely a solution with small interim storage capacity with allowing the higher number of canisters. Again, the reference point is unachievable. In Figure 6a, the reference point 4 and the solutions with $q = 4$ (r4q4) and $q = 9$ (r4q9) are illustrated. The solutions are shown in green and orange, respectively. Again, the corresponding disposal schedules are given in Figures 6b,c. As we see, the solution r4q4 satisfies the wishes towards the interim storage capacity as well as the utilization of the empty canister positions quite well. Additionally, the better values than the reference are obtained in the repository area related goals and the costs. The solution r4q9 express this as well, but the original wishes towards the interim storage capacity are not satisfied. Since the solution r4q4 captures better the ideas of the decision-maker, it is selected for the current solution f_4.

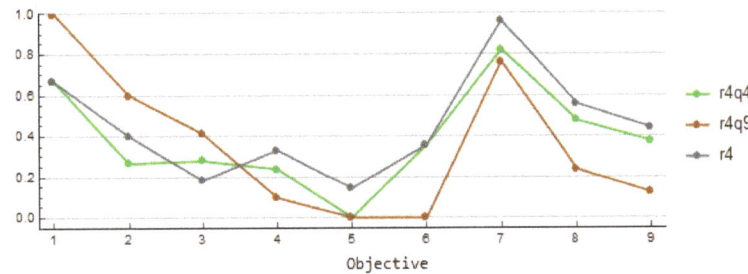

(a) The objective values for the selected solutions $q = 4$ and $q = 9$ and reference point 4.

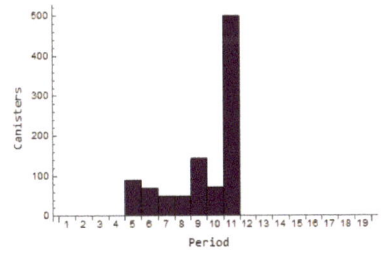

(b) The disposal schedules for $q = 4$.

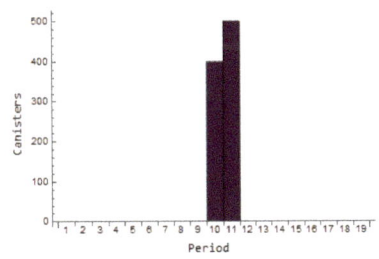

(c) The disposal schedules for $q = 9$.

Figure 6. Results for the iteration 4.

Eventually, the decision maker is ready to make the final choice. During the solution process, we have learned that the solutions obtained from 4 different reference points can be split broadly into two main groups. The first group includes solutions where the disposal starts early while the other group includes the solutions with late starting. The most striking fact is that the solutions of the first group are obtained with smaller values of q and the solutions of the second group with the larger values of q. This phenomenon is repeated with all of four reference points. Interestingly, with the modified reference point 3 where only eight objectives were considered, mainly solutions with earlier starting time were obtained. In general, the earlier starting time of the disposal improves the objectives (1)–(3) and impairs others compared with the case where the disposal starts later. In general, we notice that the solutions obtained adapt the reference points quite well.

In Figure 7, the solutions related to the first group with an early starting time are illustrated. It can be seen that even if all these solutions suggest the early start of the disposal, they still have some differences. One can improve goals (7) and (8) by disposing of spent fuel with a small volume at the beginning. However, this declines goals (1)–(3), (5), (6) and (9) which can be seen from the solution r3q2. It is possible to improve the goals (1)–(3), (5) and (6) by allowing some canister positions to be empty. However, this in its turn declines goals (4) and (7)–(9) which can be seen from the solution r4q4. As the final solution, the decision maker likes to return to the reference point 2 and the solution r2q1 looks like a good compromise when disposal begins early.

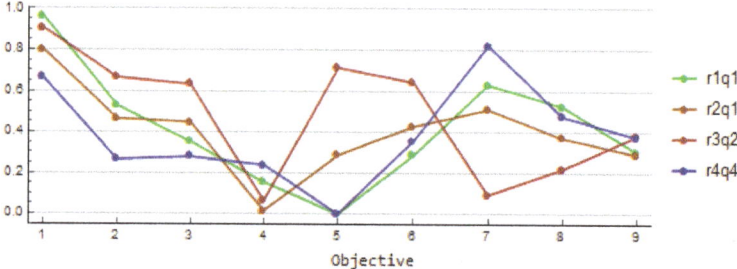

Figure 7. The four solutions where disposal starts early.

A similar examination is done for the solutions of the second group with the late starting time. The solutions in terms of the objective function values and the disposal schedules are given in Figure 8. Again, we can observe some differences. The differences depend on the number of years the start of disposal operations is prolonged. It can be seen from Figure 8, that the disposal volume is large in every solution where disposal starts late. On the one hand, one can improve goals (7)–(9) by delaying the start of disposal but on the other hand, this declines goals (2), (3) and (5), as illustrated in the solution r3q8. When the disposal starts late, empty canister positions have only a minor impact on the solution. One can improve goals (2), (3), and (5) by allowing empty canister positions. This yields to the impairing of the goals (4), and (7)–(9) which can be seen from the solution r4q9. Again, the decision maker is willing to return to the reference point 2 and consider the solution r2q9 as a satisfactory solution when the disposal starts late. Additionally, the decision maker selects this solution also for the final solution f^*, since it yields a rather good solution for other objectives than the maximum storage. However, we learned that this is the price of the lower costs and smaller disposal facility area. Moreover, compared with the solution r2q1 also presented from the reference point 2, the later starting does not delay the ending of the disposal.

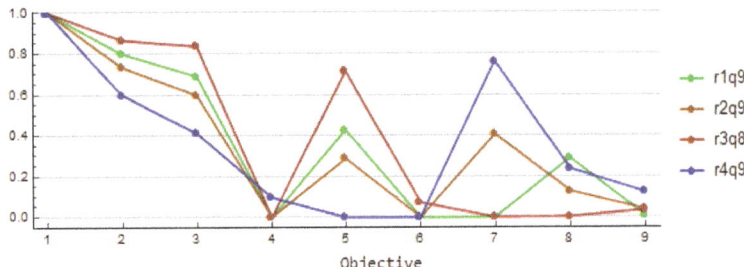

Figure 8. The four solutions where disposal starts late.

5. Conclusions

In this paper, we have proposed the nonsmooth multiobjective MINLP model to optimize the spent nuclear fuel disposal in order to obtain a disposal schedule. The modeled process is described and the model is presented in detail. Then, the two-slope parameterized ASF is briefly stated and validated the use of it. Additionally, we proposed an interactive solution method utilizing these ASFs. Finally, some numerical results from the case study are given. The solutions obtained are exemplified and analyzed in terms of objective function values and disposal schedules.

With slight modifications, the model presented is applicable to other countries than Finland as well, if the spent nuclear fuel is decided to dispose of the disposal facility. It is possible to change the objectives or leave some of them out. Indeed, this model has quite many objectives, and in some cases it may be advantageous to have fewer goals either to ease the decision maker's task or reduce the computations needed.

The schedules obtained are realistic and viable. One conspicuous feature for the solutions is that they are segmented in two groups based on the value of the parameter q enabling the parameterization when the two-slope parameterized ASF is used. With the lower values of q (i.e., closer to L_∞ metric), the disposal starts early and with the larger values of q (i.e., closer to L_1 metric) the later start of the disposal is suggested. If only one metric, for instance L_∞ metric, was used, no solutions with late starting would have been obtained in these iterations. For further studies, it would be interesting to investigate, is this kind of phenomenon observable in other applications as well, if the two-slope parameterized ASF is used. The role of q is also fascinating in terms of which value of q yields the most desirable solution for the decision maker.

As future research, it would also be interesting to include all of the three different fuel types used in Finland. Another interesting topic would be including the possible hiatus for the operation of the encapsulation facility in the model.

Author Contributions: Conceptualization, O.M., T.R. and M.M.M.; methodology, O.M. and T.R.; software, O.M.; validation, O.M.; formal analysis, O.M.; investigation, O.M.; resources, M.M.M.; data curation, O.M. and T.R.; writing—original draft preparation, O.M.; writing—review and editing, O.M., T.R. and M.M.M.; visualization, O.M.; supervision, M.M.M.; project administration, M.M.M.; funding acquisition, O.M., T.R. and M.M.M.

Funding: The study is financial supported by Emil Aaltonen foundation, University of Turku Graduate School UTUGS Matti programme, Academy of Finland project No. 294002, University of Turku, and Tampere University of Technology.

Acknowledgments: The authors are grateful for Yury Nikulin and Jani Huttunen about their comments during the preparation of this paper. The authors wish also to congratulate Adil Bagirov on his 60th birthday!

Conflicts of Interest: The authors declare no conflict of interest.

Appendix A. Parameters of the Case Study

The parameters used in the case study given in Section 4 are

$M_i = 360, i \in \{1,3,5,7,9,11\}$ $R = 4$ $p_{max}^{low} = 1300$
$M_i = 240, i \in \{2,4,6,8,10\}$ $U = 500$ $d_{DT}^{up} = 50$
$a = 5$ $T = 50$ $d_{DT}^{low} = 25$
$b = 6$ $Q = 350$ $d_{CA}^{up} = 15$
$K = 4$ $p_{max}^{up} = 1830$ $d_{CA}^{low} = 6$

and the values for $A_{i,j}$ and $P_{i,j}$ are given in Tables A1 and A2, respectively. Furthermore, the following approximation is used in the constraint (27):

$$g(p_{max}, d_{DT}) = \max\{ -2.26911 d_{DT} + 0.00675 p_{max} + 54.5228,$$
$$-0.05833 d_{DT} + 0.00596 p_{max} - 0.727083,$$
$$-0.14 d_{DT} + 0.17701 p_{max} - 350.651 \}.$$

Table A1. Values for the parameters $A_{i,j}$.

	$j=1$	$j=2$	$j=3$	$j=4$	$j=5$	$j=6$	$j=7$	$j=8$	$j=9$	$j=10$	$j=11$	$j=12$	$j=13$	$j=14$	$j=15$	$j=16$	$j=17$	$j=18$	$j=19$
$i=1$	4	5	6	7	8	9	10	11	12	13	14	15	16	17	18	19	20	21	22
$i=2$	3	4	5	6	7	8	9	10	11	12	13	14	15	16	17	18	19	20	21
$i=3$	2	3	4	5	6	7	8	9	10	11	12	13	14	15	16	17	18	19	20
$i=4$	1	2	3	4	5	6	7	8	9	10	11	12	13	14	15	16	17	18	19
$i=5$	0	1	2	3	4	5	6	7	8	9	10	11	12	13	14	15	16	17	18
$i=6$	−1	0	1	2	3	4	5	6	7	8	9	10	11	12	13	14	15	16	17
$i=7$	−2	−1	0	1	2	3	4	5	6	7	8	9	10	11	12	13	14	15	16
$i=8$	−3	−2	−1	0	1	2	3	4	5	6	7	8	9	10	11	12	13	14	15
$i=9$	−4	−3	−2	−1	0	1	2	3	4	5	6	7	8	9	10	11	12	13	14
$i=10$	−5	−4	−3	−2	−1	0	1	2	3	4	5	6	7	8	9	10	11	12	13
$i=11$	−6	−5	−4	−3	−2	−1	0	1	2	3	4	5	6	7	8	9	10	11	12

Table A2. Values for the parameters $P_{i,j}$.

	$j=1$	$j=2$	$j=3$	$j=4$	$j=5$	$j=6$	$j=7$	$j=8$	$j=9$	$j=10$	$j=11$	$j=12$	$j=13$	$j=14$	$j=15$	$j=16$	$j=17$	$j=18$	$j=19$
$i=1$	695	632	578	531	489	451	418	388	361	338	316	297	280	264	251	238	227	216	207
$i=2$	inf	695	632	578	531	489	451	418	388	361	338	316	297	280	264	251	238	227	216
$i=3$	inf	inf	695	632	578	531	489	451	418	388	361	338	316	297	280	264	251	238	227
$i=4$	inf	inf	inf	695	632	578	531	489	451	418	388	361	338	316	297	280	264	251	238
$i=5$	inf	inf	inf	inf	695	632	578	531	489	451	418	388	361	338	316	297	280	264	251
$i=6$	inf	inf	inf	inf	inf	695	632	578	531	489	451	418	388	361	338	316	297	280	264
$i=7$	inf	inf	inf	inf	inf	inf	695	632	578	531	489	451	418	388	361	338	316	297	280
$i=8$	inf	inf	inf	inf	inf	inf	inf	695	632	578	531	489	451	418	388	361	338	316	297
$i=9$	inf	inf	inf	inf	inf	inf	inf	inf	695	632	578	531	489	451	418	388	361	338	316
$i=10$	inf	inf	inf	inf	inf	inf	inf	inf	inf	695	632	578	531	489	451	418	388	361	338
$i=11$	inf	inf	inf	inf	inf	inf	inf	inf	inf	inf	695	632	578	531	489	451	418	388	361

References

1. IAEA. *The Long Term Storage of Radioactive Waste: Safety and Sustainability. A Position Paper of International Experts IAEA-LTS/RW*; IAEA: Vienna, Austria, 2003.
2. Posiva Oy. General Time Schedule for Final Disposal. Available online: http://www.posiva.fi/en/final_disposal/general_time_schedule_for_final_disposal#.XNUcxIpS-Uk (accessed on 10 May 2019).
3. Taji, K.; Levy, J.K.; Hartmann, J.; Bell, M.L.; Anderson, R.M.; Hobbs, B.F.; Feglar, T. Identifying potential repositories for radioactive waste: Multiple criteria decision analysis and critical infrastructure systems. *Int. J. Crit. Infrastruct.* **2005**, *1*, 404–422. [CrossRef]
4. Alumur, S.; Kara, B.Y. A new model for the hazardous waste location-routing problem. *Comput. Oper. Res.* **2007**, *34*, 1406–1423. [CrossRef]
5. ReVelle, C.; Cohon, J.; Shobrys, D. Simultaneous siting and routing in the disposal of hazardous wastes. *Transp. Sci.* **1991**, *25*, 138–145. [CrossRef]
6. Johnson, B.L.; Porter, A.T.; King, J.C.; Newman, A.M. Optimally configuring a measurement system to detect diversions from a nuclear fuel cycle. *Ann. Oper. Res.* **2019**, *275*, 393–420. [CrossRef]

7. Shugart, N.; Johnson, B.; King, J.; Newman, A. Optimizing nuclear material accounting and measurement systems. *Nucl. Technol.* **2018**, *204*, 260–282. [CrossRef]
8. Tosoni, E.; Salo, A.; Govaerts, J.; Zio, E. Comprehensiveness of scenarios in the safety assessment of nuclear waste repositories. *Reliab. Eng. Syst. Saf.* **2019**, *188*, 561–573. [CrossRef]
9. Ranta, T. Optimization in the Final Disposal of Spent Nuclear Fuel. Ph.D. Thesis, Tampere University of Technology, Tampere, Finland, 2012.
10. Rautman, C.A.; Reid, R.A.; Ryder, E.E. Scheduling the disposal of nuclear waste material in a geologic repository using the transportation model. *Oper. Res.* **1993**, *41*, 459–469.
11. Johnson, B.; Newman, A.; King, J. Optimizing high-level nuclear waste disposal within a deep geologic repository. *Ann. Oper. Res.* **2017**, *253*, 733–755. [CrossRef]
12. Ranta, T.; Cameron, F. Heuristic methods for assigning spent nuclear fuel assemblies to canisters for final disposal. *Nucl. Sci. Eng.* **2012**, *171*, 41–51. [CrossRef]
13. Žerovnik, G.; Snoj, L.; Ravnik, M. Optimization of spent nuclear fuel filling in canisters for deep repository. *Nucl. Sci. Eng.* **2009**, *163*, 183–190. [CrossRef]
14. Vlassopoulos, E.; Volmert, B.; Pautz, A. Logistics optimization code for spent fuel assembly loading into final disposal canisters. *Nucl. Eng. Des.* **2017**, *325*, 246–255. [CrossRef]
15. Bagirov, A.; Karmitsa, N.; Mäkelä, M.M. *Introduction to Nonsmooth Optimization: Theory, Practice and Software*; Springer: Cham, Switzerland, 2014.
16. Clarke, F.H. *Optimization and Nonsmooth Analysis*; John Wiley & Sons, Inc.: New York, NY, USA, 1983.
17. Karmitsa, N.; Bagirov, A.; Mäkelä, M.M. Comparing different nonsmooth minimization methods and software. *Optim. Methods Softw.* **2012**, *27*, 131–153. [CrossRef]
18. Bagirov, A.M.; Churilov, L. An Optimization-Based Approach to Patient Grouping for Acute Healthcare in Australia. In *Computational Science—ICCS 2003*; Sloot, P.M.A., Abramson, D., Bogdanov, A.V., Gorbachev, Y.E., Dongarra, J.J., Zomaya, A.Y., Eds.; Springer: Berlin/Heidelberg, Germany, 2003; pp. 20–29.
19. Bagirov, A.M.; Mahmood, A. A comparative assessment of models to predict monthly rainfall in Australia. *Water Resour. Manag.* **2018**, *32*, 1777–1794. [CrossRef]
20. Mäkelä, M.M.; Neittaanmäki, P. *Nonsmooth Optimization: Analysis and Algorithms with Applications to Optimal Control*; World Scientific Publishing Co.: Singapore, 1992.
21. Handl, J.; Kell, D.B.; Knowles, J. Multiobjective optimization in bioinformatics and computational biology. *IEEE/ACM Trans. Comput. Biol. Bioinform.* **2007**, *4*, 279–292. [CrossRef] [PubMed]
22. Mala-Jetmarova, H.; Barton, A.; Bagirov, A. Sensitivity of algorithm parameters and objective function scaling in multi-objective optimisation of water distribution systems. *J. Hydroinform.* **2015**, *17*, 891–916. [CrossRef]
23. Mala-Jetmarova, H.; Barton, A.; Bagirov, A. Exploration of the trade-offs between water quality and pumping costs in optimal operation of regional multiquality water distribution systems. *J. Water Resour. Plan. Manag.* **2015**, *141*, 4014077. [CrossRef]
24. Marler, R.; Arora, J. Survey of multi-objective optimization methods for engineering. *Struct. Multidiscip. Optim.* **2004**, *26*, 369–395. [CrossRef]
25. Wilppu, O.; Mäkelä, M.M.; Nikulin, Y. New Two-Slope Parameterized Achievement Scalarizing Functions for Nonlinear Multiobjective Optimization. In *Operations Research, Engineering, and Cyber Security*; Daras, N.J., Rassias, T.M., Eds.; Springer: Berlin/Heidelberg, Germany, 2017; Volume 113, pp. 403–422.
26. Nikulin, Y.; Miettinen, K.; Mäkelä, M.M. A new achievement scalarizing function based on parameterization in multiobjective optimization. *OR Spectr.* **2012**, *34*, 69–87. [CrossRef]
27. Luque, M.; Miettinen, K.; Ruiz, A.B.; Ruiz, F. A two-slope achievement scalarizing dunction for interactive multiobjective optimization. *Comput. Oper. Res.* **2012**, *39*, 1673–1681. [CrossRef]
28. Buchanan, J.; Gardiner, L. A comparison of two reference point methods in multiple objective mathematical programming. *Eur. J. Oper. Res.* **2003**, *149*, 17–34. [CrossRef]
29. Miettinen, K.; Mäkelä, M.M. On scalarizing functions in multiobjective optimization. *OR Spectr.* **2002**, *24*, 193–213. [CrossRef]
30. Miettinen, K.; Mäkelä, M.M. Synchronous approach in interactive multiobjective optimization. *Eur. J. Oper. Res.* **2006**, *170*, 909–922. [CrossRef]
31. Ehrgott, M. *Multicriteria Optimization*, 2nd ed.; Springer: Berlin/Heidelberg, Germany, 2005.
32. Miettinen, K. *Nonlinear Multiobjective Optimization*; Kluwer Academic Publishers: Boston, MA, USA, 1999.

33. Miettinen, K.; Hakanen, J.; Podkopaev, D. Interactive Nonlinear Multiobjective Optimization Methods. In *Multiple Criteria Decision Analysis: State of the Art Surveys*; Greco, S., Ehrgott, M., Figueira, J.R., Eds.; Springer: New York, NY, USA, 2016; pp. 927–976.
34. Buchanan, J.T. A naive approach for solving MCDM problems: The GUESS method. *J. Oper. Res. Soc.* **1997**, *48*, 202–206. [CrossRef]
35. Jaszkiewicz, A.; Słowiński, R. The 'Light Beam Search' approach—An overview of methodology applications. *Eur. J. Oper. Res.* **1999**, *113*, 300–314. [CrossRef]
36. Nakayama, H.; Sawaragi, Y. Satisficing Trade-off Method for Multiobjective Programming. In *Interactive Decision Analysis*; Grauer, M., Wierzbicki, A.P., Eds.; Springer: Berlin/Heidelberg, Germany, 1984; pp. 113–122.
37. Vanderpooten, D. The interactive approach in MCDA: A technical framework and some basic conceptions. *Math. Comput. Model.* **1989**, *12*, 1213–1220. [CrossRef]
38. Wierzbicki, A.P. A mathematical basis for satisficing decision making. *Math. Model.* **1982**, *3*, 391–405. [CrossRef]
39. Désidéri, J.A. Multiple-gradient descent algorithm (MGDA) for multiobjective optimization. *Compte Rendus De L'Académie Des Sci. Ser. I* **2012**, *350*, 313–318. [CrossRef]
40. Mäkelä, M.M.; Karmitsa, N.; Wilppu, O. Proximal Bundle Method for Nonsmooth and Nonconvex Multiobjective Optimization. In *Mathematical Modeling and Optimization of Complex Structures*; Tuovinen, T., Repin, S., Neittaanmäki, P., Eds.; Springer: Berlin/Heidelberg, Germany, 2016; Volume 40, pp. 191–204.
41. Montonen, O.; Karmitsa, N.; Mäkelä, M.M. Multiple subgradient descent bundle method for convex nonsmooth multiobjective optimization. *Optimization* **2018**, *67*, 139–158. [CrossRef]
42. Montonen, O.; Joki, K. Bundle-based descent method for nonsmooth multiobjective DC optimization with inequality constraints. *J. Glob. Optim.* **2018**, *72*, 403–429. [CrossRef]
43. Qu, S.; Liu, C.; Goh, M.; Li, Y.; Ji, Y. Nonsmooth multiobjective programming with quasi-Newton methods. *Eur. J. Oper. Res.* **2014**, *235*, 503–510. [CrossRef]
44. Kilinc, M.R.; Sahinidis, N.V. Exploiting integrality in the global optimization of mixed-integer nonlinear programming problems with BARON. *Optim. Methods Softw.* **2018**, *33*, 540–562. [CrossRef]
45. Tawarmalani, M.; Sahinidis, N.V. A polyhedral branch-and-cut approach to global optimization. *Math. Program.* **2005**, *103*, 225–249. [CrossRef]
46. GAMS Development Corporation. General Algebraic Modeling System (GAMS) Release 26.1.0. Washington, DC, USA. Available online: http://www.gams.com/ (accessed on 16 April 2019).

© 2019 by the authors. Licensee MDPI, Basel, Switzerland. This article is an open access article distributed under the terms and conditions of the Creative Commons Attribution (CC BY) license (http://creativecommons.org/licenses/by/4.0/).

Article

SVM-Based Multiple Instance Classification via DC Optimization

Annabella Astorino [1,*,†], Antonio Fuduli [2,†], Giovanni Giallombardo [3,†] and Giovanna Miglionico [3,†]

1   Istituto di Calcolo e Reti ad Alte Prestazioni-C.N.R., 87036 Rende (CS), Italy
2   Dipartimento di Matematica e Informatica, Università della Calabria, 87036 Rende (CS), Italy; antonio.fuduli@unical.it
3   Dipartimento di Ingegneria Informatica, Modellistica, Elettronica e Sistemistica, Università della Calabria, 87036 Rende (CS), Italy; giovanni.giallombardo@unical.it (G.G.); gmiglionico@dimes.unical.it (G.M.)
*   Correspondence: annabella.astorino@icar.cnr.it
†   These authors contributed equally to this work.

Received: 31 October 2019; Accepted: 20 November 2019; Published: 23 November 2019

Abstract: A multiple instance learning problem consists of categorizing objects, each represented as a set (bag) of points. Unlike the supervised classification paradigm, where each point of the training set is labeled, the labels are only associated with bags, while the labels of the points inside the bags are unknown. We focus on the binary classification case, where the objective is to discriminate between positive and negative bags using a separating surface. Adopting a support vector machine setting at the training level, the problem of minimizing the classification-error function can be formulated as a nonconvex nonsmooth unconstrained program. We propose a difference-of-convex (DC) decomposition of the nonconvex function, which we face using an appropriate nonsmooth DC algorithm. Some of the numerical results on benchmark data sets are reported.

Keywords: multiple instance learning; support vector machine; DC optimization; nonsmooth optimization

1. Introduction

Multiple instance learning (MIL) is a recent machine learning paradigm [1–3], which consists of classifying sets of points. Each set is called bag, while the points inside the bags are called instances. The main characteristic of an MIL problem is that in the learning phase the instance labels are hidden and only the labels of the bags are known.

An MIL seminal paper is [4], where a drug-design problem has been faced. Such a problem consists of determining whether a drug molecule (bag) is active or non-active. A molecule provides the desired drug effect (positive label) if, and only if, at least one of its conformations (instances) binds to the target site. The crucial question is that it is not known a priori which conformation makes the molecule active.

Some MIL applications are image classification [5–8], drug discovery [9,10], classification of text documents [11], bankruptcy prediction [12], and speaker identification [13].

For this kind of problems, there are various solutions in the literature that fall into three different classes: instance-space approaches, bag-space approaches, and embedding-space approaches. In instance-space approaches, classification is performed at the instance level, finding a separation surface directly in the instance space, without looking at the global structure of the bags; the label of each bag is determined as an aggregation of the labels of its corresponding instances. Vice-versa, in bag-space approaches (for example, see [14–16]), the separation is performed at a global level, considering the bag as a whole entity. A compromise between these two kinds of approaches is constituted by embedding-space techniques, where each bag is represented by one feature vector and the classification is consequently performed in the instance space. An example of an embedding-space approach is presented in [17].

The method we propose uses the instance-space approach and provides a separation hyperplane for the binary case, where the objective is to discriminate between positive and negative bags. We start from the standard MIL assumption stating that a bag is positive if, and only if, at least one of its instances is positive and it is negative whenever all its instances are negative.

Some examples of linear instance-space MIL classifiers can be found in [18–22]. In particular, in [18], two different models have been proposed. The first one is a mixed-integer nonlinear optimization problem solved using a heuristic technique based on the block coordinate descent method [23] and faced in [19] using a Lagrangian relaxation technique. The second model, which will be the objective of our analysis in the next section, is a nonsmooth nonconvex optimization problem, solved in [21] using the bundle type method described in [24]. In [20], a semi-proximal support vector machine (SVM) approach is used, coming from a compromise between the classical SVM [25] and the proximal approach proposed in [26] for supervised classification. Finally, an optimization problem with bilinear constraints is analyzed in [22], where each positive bag is expressed as a convex combination of its instances and a local solution is obtained by solving successive linear programs.

Recently, nonlinear instance-space MIL classifiers have also been proposed in the literature, such as in [27] and in [28], where a spherical separation approach is adopted: in particular, in the former a variable neighborhood search method [29] is used, while in the latter a DC (difference of convex) model is solved using an appropriate DC algorithm [30]. In passing, we stress that many DC models have been introduced in machine learning, in the supervised [31–35], semisupervised [36,37] and unsupervised cases [38–40].

In this work, we propose a DC optimization model providing a linear classifier for binary MIL problems. The solution method we adopt is the proximal bundle method introduced in [30] for the minimization of nonsmooth DC functions. The paper is organized as follows. In the next two sections, we describe, respectively, the DC optimization model and the corresponding nonsmooth solution algorithm. Finally, in Section 4, we report the results of our numerical experimentation performed on some data sets drawn from the literature.

2. A DC Decomposition of the SVM-Based MIL

We tackle a binary MIL problem whose objective is to discriminate between m positive bags and k negative ones using a hyperplane

$$H(w,b) \triangleq \{x \in \mathbb{R}^n \mid w^T x + b = 0\},$$

where $w \in \mathbb{R}^n$ and $b \in \mathbb{R}$. Indicating by J_i^+, $i = 1, \ldots, m$, the index set of the instances belonging to the ith positive bag and by J_i^-, $i = 1, \ldots, k$, the index set of the instances belonging to the ith negative bag, we recall that, on the basis of the standard MIL assumption, a bag is positive if, and only if, at least one of its instances is positive and it is negative vice-versa. As a consequence, while a positive bag is allowed to, possibly, straddle the hyperplane, the negative bags should lie completely on the negative side.

More formally, indicating by $x_j \in \mathbb{R}^n$ the jth instance of a positive or negative bag, the hyperplane $H(w,b)$ performs a correct separation if, and only if, the following conditions hold:

$$\begin{cases} w^T x_j + b \geq 1, & \text{for at least an index } j \in J_i^+ \text{ and for all } i = 1, \ldots, m \\ w^T x_j + b \leq -1, & \text{for all } j \in J_i^- \text{ and for all } i = 1, \ldots, k. \end{cases}$$

As a consequence (see Figures 1 and 2), a positive bag J_i^+, $i = 1, \ldots, m$, is misclassified if

$$\max_{j \in J_i^+}(w^T x_j + b) < 1$$

and a negative one J_i^-, $i = 1, \ldots, k$, is misclassified if

$$\max_{j \in J_i^-}(w^T x_j + b) > -1.$$

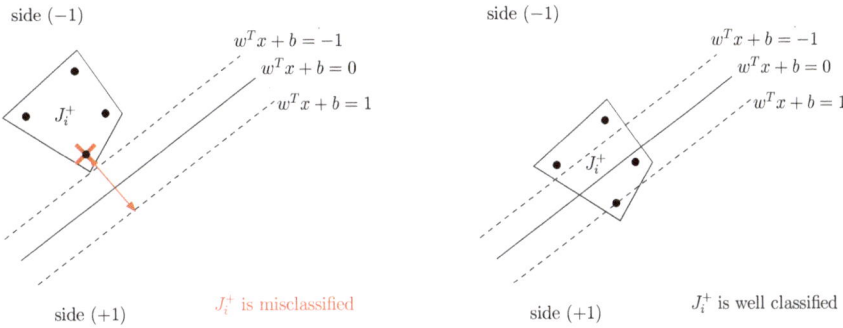

Figure 1. Positive bag J_i^+.

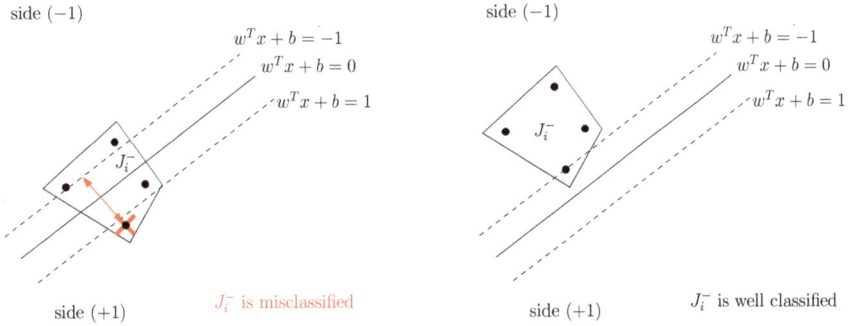

Figure 2. Negative bag J_i^-.

Then, we come out with the following error function, already introduced in [18]:

$$f(w,b) \triangleq \frac{1}{2}\|w\|^2 + C \left[\sum_{i=1}^{m} \max\{0, 1 - \max_{j \in J_i^+}(w^T x_j + b)\} + \sum_{i=1}^{k} \max\{0, 1 + \max_{j \in J_i^-}(w^T x_j + b)\} \right], \quad (1)$$

where $C > 0$ represents the trade-off between two objectives: the maximization of the separation margin, characterizing the classical SVM [25] approach, and the minimization of the classification error. To minimize function f, we propose a DC decomposition based on the following formula:

$$\max\{0, 1 - h(y)\} = \max\{1, h(y)\} - h(y), \quad (2)$$

where h is a convex function. By applying Equation (2) to our case, we can write f in the form:

$$f(w,b) = f_1(w,b) - f_2(w,b),$$

where

$$f_1(w,b) \triangleq \frac{1}{2}\|w\|^2 + C\sum_{i=1}^{k}\max\{0, 1 + \max_{j \in J_i^-}(w^T x_j + b)\} + C\sum_{i=1}^{m}\max\{1, \max_{j \in J_i^+}(w^T x_j + b)\}$$

and

$$f_2(w,b) \triangleq C\sum_{i=1}^{m}\max_{j \in J_i^+}(w^T x_j + b)$$

are convex functions. Hence, we come up with the following nonconvex nonsmooth optimization problem, DC-MIL,

$$\min_{w,b}[f_1(w,b) - f_2(w,b)]. \tag{3}$$

3. Solving DC-MIL using a Nonsmooth DC Algorithm

We start by recalling some preliminary property of the DC optimization problem, by adopting the same notation as above. Given the DC optimization problem

$$\min_{y}[f_1(y) - f_2(y)] \tag{4}$$

where both f_1 and f_2 are convex nonsmooth functions, we say that a point y^* is a local minimizer if $f_1(y^*) - f_2(y^*)$ is finite and there exists a neighborhood $\mathcal{N}$ of y^* such that

$$f_1(y^*) - f_2(y^*) \leq f_1(y) - f_2(y), \quad \forall y \in \mathcal{N}. \tag{5}$$

Considering that, in general, the Clarke subdifferential calculus cannot be used to compute subgradients of the DC function since

$$\partial_{cl} f(y) \subseteq \partial f_1(y) - \partial f_2(y), \tag{6}$$

where $\partial_{cl} f(\cdot)$ denotes Clarke's subdifferential, different stationary points can be defined for nonsmooth DC functions. A point y^* is called inf-stationary for problem Equation (4) if

$$\emptyset \neq \partial f_2(y^*) \subseteq \partial f_1(y^*). \tag{7}$$

Furthermore, a point y^* is called Clarke stationary for problem Equation (4) if

$$0 \in \partial_{cl} f(y^*), \tag{8}$$

while, it is called a critical point of f if

$$\partial f_2(y^*) \cap \partial f_1(y^*) \neq \emptyset. \tag{9}$$

Denoting the set of inf-stationary points by S_{inf}, the set of Clarke stationary points by S_{cl}, and the set of critical points of the function f by S_{cr}, the following inclusions hold

$$S_{inf} \subseteq S_{cl} \subseteq S_{cr}$$

as shown in (Proposition 3, [30]).

Nonsmooth DC functions have attracted the interest of several researchers, both, from the theoretical and from the algorithmic viewpoint. Focusing in particular on the algorithmic side, the most relevant contribution has probably been provided by the methods based on the linearization of function f_2 (see, [41] and references therein), where the problem is tackled via successive convexifications of function f. In the last years, nonsmooth-tailored DC programming has experienced a lot of attention as it has a lot of practical applications (see [28,42]). In fact, several nonsmooth DC algorithms have been developed ([30,43–47]).

Here, we adopt the algorithm DCPCA, a bundle-type method introduced in [30] to solve problem Equation (4), which is based on a model function built by combining two convex piecewise approximations, each related to one component function. More in details, a simplified version of DCPCA works as follows:

- It iteratively builds two separate piecewise-affine approximations of the component functions, grouping the corresponding information in two separate bundles.
- It combines the two convex piecewise-affine approximations and generates a DC piecewise-affine model.
- The DC (hence, nonconvex) model is locally approximated using an auxiliary quadratic program, whose solution is used to certify approximate criticality, or to generate a descent search-direction to be explored via a backtracking line-search approach.
- Whenever no descent is achieved along the search direction, the bundle of the first function is enriched, thus, obtaining a better model function with this being the fundamental feature of any cutting plane algorithm.

In fact, the DCPCA is based on constructing a model function as the pointwise maximum of several concave piecewise-affine pieces. To construct this model, starting from some cutting-plane ideas, the information coming from the two component functions are kept separate in two bundles. We denote the stability center by z (i.e., an estimate of the minimizer), and by I and L, the index sets of the points generated by the algorithm where the information of function f_1 and f_2 have been evaluated, respectively. Therefore, we denote the two bundles of information as

$$\mathcal{B}_1 = \{(g_i^{(1)}, \alpha_i^{(1)}) : i \in I\}$$

and

$$\mathcal{B}_2 = \{(g_l^{(2)}, \alpha_l^{(2)}) : l \in L\}$$

where, for every $i \in I$, $g_i^{(1)} \in \partial f_1(y_i)$ with

$$\alpha_i^{(1)} = f_1(z) - \left(f_1(y_i) + g_i^{(1)T}(z - y_i)\right),$$

and, for every $l \in L$, $g_l^{(2)} \in \partial f_2(y_l)$ with

$$\alpha_l^{(2)} = f_2(z) - \left(f_2(y_l) + g_l^{(2)T}(z - y_l)\right).$$

We remark that both component functions, along with their subgradients, could be evaluated at some iterate-point, and, indeed, we assume that $(g^{(1)}(z), 0) \in \mathcal{B}_1$ and $(g^{(2)}(z), 0) \in \mathcal{B}_2$, where $g^{(1)}(z) \in \partial f_1(z)$ and $g^{(2)}(z) \in \partial f_2(z)$.

To approximate the difference function

$$\left(f_1(z+d) - f_2(z+d)\right) - \left(f_1(z) - f_2(z)\right)$$

at a given iteration k the following nonconvex model function $\Gamma_k(d)$ is introduced

$$\Gamma_k(d) \triangleq \max_{i \in I} \min_{l \in L} \left\{ (g_i^{(1)} - g_l^{(2)})^T d - \alpha_i^{(1)} + \alpha_l^{(2)} \right\}, \tag{10}$$

which is defined as the maximum of finitely many concave piecewise-affine functions. The model-function Γ_k is used to state a sufficient descent condition of the type

$$\left(f_1(z+d) - f_2(z+d) \right) - \left(f_1(z) - f_2(z) \right) \leq m\Gamma_k(d)$$

where $m \in (0,1)$. The interesting property of such a model-function is that whenever the sufficient descent is not achieved at points that are close to the stability center, say $z + \bar{d}$, then an improved cutting-plane model can be obtained by only updating the bundle of f_1 with the appropriate information related to the point $z + \bar{d}$. On the other hand, it looks obviously difficult to adopt the minimization of the model-function Γ_k as a building block of any algorithm, given its nonconvexity. In fact, DCPCA does not require the direct minimization of $\Gamma_k(d)$, but the search direction can be obtained by solving the following auxiliary quadratic problem:

$$\min_{d \in \mathbb{R}^n, v \in \mathbb{R}} \quad v + \frac{1}{2}\|d\|^2$$
$$v \geq (g_i^{(1)} - g_{\bar{l}}^{(2)})^T d - \alpha_i^{(1)} \quad \forall i \in I \qquad QP(I)$$

where $\bar{l} \in L(0) \triangleq \{l \in L : \alpha_l^{(2)} = 0\}$. We observe that $L(0) \neq \emptyset$ as $\mathcal{B}_2$ is assumed to contain the information about the current stability center. More precisely, DCPCA works by forcing $L(0)$ to be a singleton, hence by letting $g_{\bar{l}}^{(2)} = g^{(2)}(z)$. Denoting the unique optimal solution of Equation ($QP(I)$) by $(\bar{d}, \bar{v})$, a standard duality argument ensures that

$$\bar{d} = -\sum_{i \in I} \bar{\lambda}_i (g_i^{(1)} - g_{\bar{l}}^{(2)}) \tag{11}$$

$$\bar{v} = -\left\| \sum_{i \in I} \bar{\lambda}_i (g_i^{(1)} - g_{\bar{l}}^{(2)}) \right\|^2 - \sum_{i \in I} \bar{\lambda}_i \alpha_i^{(1)} \tag{12}$$

where $\bar{\lambda}_i \geq 0$, $i \in I$, are the optimal variables of the dual of $QP(I)$, with $\sum_{i \in I} \bar{\lambda}_i = 1$.

Given that any starting point $z = y_0$, DCPCA returns an approximate critical point z^*, see (Theorem 1, [30]). The following parameters are adopted: the optimality parameter $\theta > 0$, the subgradient threshold $\eta > 0$, the linearization-error threshold $\varepsilon > 0$, the approximate line-search parameter $m \in (0,1)$, and the step-size reduction parameter $\sigma \in (0,1)$. In Algorithm 1, we report an algorithmic scheme of the main iteration, namely of the set of steps where the stability center is unchanged. An exit from the main iteration is obtained as soon as a stopping criterion is satisfied or whenever the stability center is updated. To make the presentation clearer, without impairing convergence properties, we skip the description of some rather technical steps, which are strictly related to the management of bundle $\mathcal{B}_2$. Details can be found in [30].

Algorithm 1 DCPCA Main Iteration

1: Solve $QP(I)$ and obtain $(\bar{d}, \bar{v})$ ▷ Find the search-direction and the predicted-reduction
2: **if** $|\bar{v}| \leq \theta$ **then** ▷ Stopping test
3:   set $z^* = z$ and **exit** ▷ Return the approximate critical point z^*
4: **end if**
5: Set $t = 1$ ▷ Start the line-search
6: **if** $f(z + t\bar{d}) - f(z) \leq mt\bar{v}$ **then** ▷ Descent test
7:   set $z := z + t\bar{d}$ ▷ Make a serious step
8:   calculate $g_+^{(1)} \in \partial f_1(z)$ and $g_+^{(2)} \in \partial f_2(z)$ ▷
9:   update $\alpha_i^{(1)}$ for all $i \in I$ and $\alpha_l^{(2)}$ for all $l \in L$ ▷
10: set $\mathcal{B}_1 = \mathcal{B}_1 \setminus \{(g_i^{(1)}, \alpha_i^{(1)}) : \alpha_i^{(1)} > \varepsilon, i \in I\} \cup \{(g_+^{(1)}, 0)\}$ ▷
11: set $\mathcal{B}_2 = \mathcal{B}_2 \cup \{(g_+^{(2)}, 0)\}$ ▷
12: update appropriately I and L, and **go to** 1 ▷
13: **else if** $t\|\bar{d}\| > \eta$ **then** ▷ Closeness test
14: set $t = \sigma t$ and **go to** 6 ▷ Reduce the step-size and iterate the line-search
15: **end if**
16: Calculate $g_+^{(1)} \in \partial f_1(z + t\bar{d})$ ▷ Make a null step
17: calculate $\alpha_+^{(1)} = f_1(z) - f_1(z + t\bar{d}) + tg_+^{(1)\top}\bar{d}$ ▷
18: set $\mathcal{B}_1 = \mathcal{B}_1 \cup \{(g_+^{(1)}, \alpha_+^{(1)})\}$, update appropriately I, and **go to** 1 ▷

We remark that the stopping condition $\bar{v} \geq -\theta$, checked at Step 2 of the DCPCA, is an approximate θ-criticality condition for z^*. Indeed, taking into account Equation (12), the stopping condition ensures that

$$\left\|\sum_{i \in I} \lambda_i^* g_i^{(1)} - g_l^{(2)}\right\| \leq \sqrt{\theta} \quad \text{and} \quad \left\|\sum_{i \in I} \lambda_i^* \alpha_i^{(1)}\right\| \leq \sqrt{\theta},$$

which in turn implies that $g_*^{(1)} \in \partial_\theta f_1(z^*)$ and $g_*^{(2)} \in \partial f_2(z^*)$ such that

$$\|g_*^{(1)} - g_*^{(2)}\|^2 \leq \theta,$$

namely, that

$$\text{dist}\left(\partial_\theta f_1(z^*), \partial f_2(z^*)\right) \leq \theta,$$

an approximate θ-criticality condition for z^*, see Equation (9).

4. Results

We tested the performance of the algorithm DCPCA applied to the DC-MIL formulation (3) by adopting two sets of medium- and large-size problems extracted from [18]. The relevant characteristics of each problem are reported in Tables 1 and 2, where we list the problem dimension n, the number of instances, and the number of bags.

Table 1. Medium-size test problems.

Data Set	Dimension	Instances	Bags
Elephant	230	1320	200
Fox	230	1320	200
Tiger	230	1220	200
Musk-1	166	476	92
Musk-2	166	6598	102

Table 2. Large-size test problems.

Data Set	Dimension	Instances	Bags
TST1	6668	3224	400
TST2	6842	3344	400
TST3	6568	3246	400
TST4	6626	3391	400
TST7	7037	3367	400
TST9	6982	3300	400
TST10	7073	3453	400

The two-level cross-validation protocol has been used to tune C and to train the classifier. Before proceeding with the training phase, the model-selection phase is aimed at finding a promising value of parameter C in the set $\{2^{-7}, 2^{-6}, \ldots, 1, \ldots, 2^6, 2^7\}$, using a lower-level cross-validation protocol on each training set. The selected C value, for each training set, is the one returning the highest average test-correctness in the model-selection phase.

Choosing a good starting point is a critical phase to ensure good performance for a local optimization algorithm like DCPCA. For each training set, denoting the barycenter of all the instances belonging to positive bags by $\overline{w}_+$ and the barycenter of all the instances belonging to negative bags by $\overline{w}_-$, we have selected the starting point (w_0, b_0) by setting

$$w_0 = \overline{w}_+ - \overline{w}_- \tag{13}$$

and choosing b_0 such that the corresponding hyperplane correctly classifies all the positive bags.

We adopted the Java implementation of algorithm DCPCA by running the computational experiments on a 3.50 GHz Intel Core i7 computer. We limited the computational budget for every execution of DCPCA to 500 and 200 evaluations of the objective function for medium-size and large-size problems, respectively, and we restricted the size of the bundle to 100 elements adopting a restart strategy, as soon as, the bundle size exceeds the threshold and a new stability center is obtained. The QP solver of IBM ILOG CPLEX 12.8 has been used to solve quadratic subprograms. The following set of parameters, according to the notation introduced in [45], has been selected: the optimality parameter $\theta = 0.7$, the subgradient threshold $\eta = 0.7$, the approximate linesearch parameter $m = 0.01$, the step-size reduction parameter $\sigma = 0.01$, and the linearization-error threshold $\epsilon = 0.95$.

We compare our DC-MIL approach against the algorithms mi-SVM [18], MI-SVM [18], MICA [22], MIL-RL [19], and for medium-size problems also against the MIC^{Bundle} [21] and DC-SMIL [28]. All such methods have been briefly surveyed in the introduction section.

To analyze the reliability of our approach, in Tables 3 and 4, we report the numerical results in terms of the percentage test-correctness averaged over 10 folds, with the best performance being underlined. We remark that some data are not reported in Table 5 as the corresponding results are obtained by adopting only nonlinear kernels in [18,22]. Moreover, to provide some insight into the efficiency of DC-MIL, we report in Tables 5 and 6, the average train-correctness (**train**, %), the average cpu time (**cpu**, sec), the average number of function evaluations (**nF**), and the average number of subgradient evaluations of the two functions (**nG1** and **nG2**). The reliability results show a good and balanced performance of the DC-MIL approach equipped with DCPCA, both, for the medium-size problems, where in one case DC-MIL slightly outperforms the other approaches, and for the large-size problems. Moreover, we observe that our approach looks strongly efficient as it manages to achieve high train-correctness in reasonably small execution times even for large-size problems.

Table 3. Average test-correctness (%) for medium-size problems.

Data Set	DC-MIL	MIL-RL	DC-SMIL	mi-SVM	MI-SVM	MICA	MIC^{Bundle}
Elephant	84.0	83.0	84.5	82.2	81.4	80.5	80.5
Fox	57.0	54.5	56.0	58.2	57.8	58.7	58.3
Tiger	84.5	75.0	81.0	78.4	84.0	82.6	79.1
Musk-1	74.5	80.0	76.7	-	-	-	75.6
Musk-2	74.0	73.0	79.0	-	-	-	76.8

Underlined means the best performance being.

Table 4. Average test-correctness (%) for large-size problems.

Data Set	DC-MIL	MIL-RL	mi-SVM	MI-SVM	MICA
TST01	94.3	95.5	93.6	93.9	94.5
TST02	80.0	85.5	78.2	84.5	85.0
TST03	86.5	86.8	87.0	82.2	86.0
TST04	86.0	79.8	82.8	82.4	87.7
TST07	79.8	83.5	81.3	78.0	78.9
TST09	68.3	68.8	67.5	60.2	61.4
TST10	78.0	77.5	79.6	79.5	82.3

Underlined means the best performance being.

Table 5. DC-MIL average efficiency. Medium-size test problems.

Data Set	Train	Cpu	nF	nG1	nG2
Elephant	91.0	3.14	500	243	208
Fox	79.9	3.05	500	81	80
Tiger	95.5	2.83	500	237	197
Musk-1	96.9	1.29	500	197	177
Musk-2	93.5	6.52	500	174	167

Table 6. DC-MIL average efficiency. Large-size test problems.

Data Set	Train	Cpu	nF	nG1	nG2
TST01	100.0	70.22	200	93	91
TST02	94.2	69.87	200	83	82
TST03	99.6	64.77	200	82	81
TST04	93.5	67.58	200	84	83
TST07	99.2	74.11	200	85	84
TST09	94.4	67.99	200	82	81
TST10	91.9	72.24	200	81	80

5. Conclusions

We have considered a multiple instance learning problem consisting of classifying sets instead of single points. The resulting binary classification problem, addressed by a support vector machine approach, is formulated as an unconstrained nonsmooth optimization problem for which an original DC decomposition is presented. The problem is solved by a proximal bundle-type method, specialized for nonsmooth DC optimization, which is tested on some benchmark datasets against a set of state-of-the-art approaches. The numerical results in terms of reliability show, on one hand, that there are no outperforming methods on all the test problems, on the other hand, that our method achieves comparable performance with other approaches. Moreover, the encouraging results obtained in terms of efficiency show that there is room for improvement by further investigating the parameter settings in relation to specific test problems.

Author Contributions: Methodology, A.A., A.F., G.G., G.M.; software, A.A., A.F., G.G., G.M.; writing–review & editing, A.A., A.F., G.G., G.M.

Funding: This research received no external funding.

Conflicts of Interest: The authors declare no conflict of interest.

Abbreviations

The following abbreviations are used in this manuscript:

MIL   Multiple instance learning
SVM   Support vector machine
DC   Difference of convex

References

1. Amores, J. Multiple instance classification: Review, taxonomy and comparative study. *Artif. Intell.* **2013**, *201*, 81–105. [CrossRef]
2. Carbonneau, M.; Cheplygina, V.; Granger, E.; Gagnon, G. Multiple instance learning: a survey of problem characteristics and applications. *Pattern Recognit.* **2018**, *77*, 329 – 353. [CrossRef]
3. Herrera, F.; Ventura, S.; Bello, R.; Cornelis, C.; Zafra, A.; Sanchez-Tarrago, D.; Vluymans, S. *Multiple Instance Learning. Foundations and Algorithms*; Springer: Berlin/Heidelberg, Germany, 2016; pp. 1–233.
4. Dietterich, T.G.; Lathrop, R.H.; Lozano-Pérez, T. Solving the multiple instance problem with axis-parallel rectangles. *Artif. Intell.* **1997**, *89*, 31–71. [CrossRef]
5. Astorino, A.; Fuduli, A.; Gaudioso, M.; Vocaturo, E. Multiple Instance Learning Algorithm for Medical Image Classification. *CEUR Workshop Proceedings* 2019. Volume 2400. Available online: http://ceur-ws.org/Vol-2400/paper-46.pdf (accessed on 25 September 2019).
6. Astorino, A.; Fuduli, A.; Veltri, P.; Vocaturo, E. Melanoma detection by means of multiple instance learning. *Interdiscip. Sci. Comput. Life Sci.* **2019**. [CrossRef] [PubMed]
7. Astorino, A.; Gaudioso, M.; Fuduli, A.; Vocaturo, E. A multiple instance learning algorithm for color images classification. In *ACM International Conference Proceeding Series*; ACM: New York, NY, USA, 2018; pp. 262–266.
8. Quellec, G.; Cazuguel, G.; Cochener, B.; Lamard, M. Multiple-Instance Learning for Medical Image and Video Analysis. *IEEE Rev. Biomed. Eng.* **2017**, *10*, 213–234. [CrossRef] [PubMed]
9. Fu, G.; Nan, X.; Liu, H.; Patel, R.Y.; Daga, P.R.; Chen, Y.; Wilkins, D.E.; Doerksen, R.J. Implementation of multiple-instance learning in drug activity prediction. *BMC Bioinform.* **2012**, *13*. [CrossRef] [PubMed]
10. Zhao, Z.; Fu, G.; Liu, S.; Elokely, K.M.; Doerksen, R.J.; Chen, Y.; Wilkins, D.E. Drug activity prediction using multiple-instance learning via joint instance and feature selection. *BMC BioInform.* **2013**, *14*. [CrossRef]
11. Liu, B.; Xiao, Y.; Hao, Z. A selective multiple instance transfer learning method for text categorization problems. *Knowl.-Based Syst.* **2018**, *141*, 178–187. [CrossRef]
12. Kotsiantis, S.; Kanellopoulos, D. Multi-instance learning for bankruptcy prediction. In Proceedings of the 2008 Third International Conference on Convergence and Hybrid Information Technology, Busan, Korea, 11–13 November 2008; Volume 1, pp. 1007–1012.
13. Briggs, F.; Lakshminarayanan, B.; Neal, L.; Fern, X.Z.; Raich, R.; Hadley, S.J.K.; Hadley, A.S.; Betts, M.G. Acoustic classification of multiple simultaneous bird species: A multi-instance multi-label approach. *J. Acoust. Soc. Am.* **2012**, *131*, 4640–4650. [CrossRef]
14. Gärtner, T.; Flach, P.A.; Kowalczyk, A.; Smola, A.J. Multi-instance kernels. In Proceedings of the 19th International Conference on Machine Learning, Sydney, Australia, 8–12 July 2002; pp. 179–186.
15. Wang, J.; Zucker, J.D. Solving the multiple-instance problem: a lazy learning approach. In Proceedings of the Seventeenth International Conference on Machine Learning, Stanford, CA, USA, 29 June–2 July 2000; Morgan Kaufmann: San Francisco, CA, USA, 2000; pp. 1119–1126.
16. Wen, C.; Zhou, M.; Li, Z. Multiple instance learning via bag space construction and ELM. In Proceedings of the International Society for Optical Engineering, Shanghai, China, 15–17 August 2018; Volume 10836.

17. Wei, X.; Wu, J.; Zhou, Z. Scalable Algorithms for Multi-Instance Learning. *IEEE Trans. Neural Netw. Learn. Syst.* **2017**, *28*, 975–987. [CrossRef]
18. Andrews, S.; Tsochantaridis, I.; Hofmann, T. Support vector machines for multiple-instance learning. In *Advances in Neural Information Processing Systems*; Becker, S., Thrun, S., Obermayer, K., Eds.; MIT Press: Cambridge, UK, 2003; pp. 561–568.
19. Astorino, A.; Fuduli, A.; Gaudioso, M. A Lagrangian relaxation approach for binary multiple instance classification. *IEEE Trans. Neural Netw. Learn. Syst.* **2019**, *30*, 2662–2671. [CrossRef] [PubMed]
20. Avolio, M.; Fuduli, A. A semi-proximal support vector machine approach for binary multiple instance learning. **2019**, submitted.
21. Bergeron, C.; Moore, G.; Zaretzki, J.; Breneman, C.; Bennett, K. Fast bundle algorithm for multiple instance learning. *IEEE Trans. Pattern Anal. Mach. Intell.* **2012**, *34*, 1068–1079. [CrossRef] [PubMed]
22. Mangasarian, O.; Wild, E. Multiple instance classification via successive linear programming. *J. Optim. Theory Appl.* **2008**, *137*, 555–568. [CrossRef]
23. Tseng, P. Convergence of a block coordinate descent method for nondifferentiable minimization. *J. Optim. Theory Appl.* **2001**, *109*, 475–494. [CrossRef]
24. Fuduli, A.; Gaudioso, M.; Giallombardo, G. Minimizing nonconvex nonsmooth functions via cutting planes and proximity control. *SIAM J. Optim.* **2004**, *14*, 743–756. [CrossRef]
25. Vapnik, V. *The Nature of the Statistical Learning Theory*; Springer: New York, NY, USA, 1995.
26. Fung, G.; Mangasarian, O. Proximal support vector machine classifiers. In Proceedings of the Seventh ACM Sigkdd International Conference on Knowledge Discovery and Data Mining, San Francisco, CA, USA, 26–29 August 2001; Provost, F., Srikant, R., Eds.; ACM: New York, NY, USA, 2001, pp. 77–86.
27. Plastria, F.; Carrizosa, E.; Gordillo, J. Multi-instance classification through spherical separation and VNS. *Comput. Oper. Res.* **2014**, *52*, 326–333. [CrossRef]
28. Gaudioso, M.; Giallombardo, G.; Miglionico, G.; Vocaturo, E. Classification in the multiple instance learning framework via spherical separation. *Soft Comput.* **2019**. [CrossRef]
29. Hansen, P.; Mladenović, N.; Moreno Pérez, J.A. Variable neighbourhood search: Methods and applications. *4OR* **2008**, *6*, 319–360. [CrossRef]
30. Gaudioso, M.; Giallombardo, G.; Miglionico, G.; Bagirov, A.M. Minimizing nonsmooth DC functions via successive DC piecewise-affine approximations. *J. Glob. Optim.* **2018**, *71*, 37–55. [CrossRef]
31. Astorino, A.; Fuduli, A.; Gaudioso, M. DC models for spherical separation. *J. Glob. Optim.* **2010**, *48*, 657–669. [CrossRef]
32. Astorino, A.; Fuduli, A.; Gaudioso, M. Margin maximization in spherical separation. *Comput. Optim. Appl.* **2012**, *53*, 301–322. [CrossRef]
33. Astorino, A.; Gaudioso, M.; Seeger, A. Conic separation of finite sets. I. The homogeneous case. *J. Convex Anal.* **2014**, *21*, 1–28.
34. Astorino, A.; Gaudioso, M.; Seeger, A. Conic separation of finite sets. II. The non-homogeneous case. *J. Convex Anal.* **2014**, *21*, 819–831.
35. Le Thi, H.A.; Le, H.M.; Pham Dinh, T.; Van Huynh, N. Binary classification via spherical separator by DC programming and DCA. *J. Glob. Optim.* **2013**, *56*, 1393–1407. [CrossRef]
36. Astorino, A.; Fuduli, A. Semisupervised spherical separation. *Appl. Math. Model.* **2015**, *39*, 6351–6358. [CrossRef]
37. Wang, J.; Shen, X.; Pan, W. On efficient large margin semisupervised learning: Method and theory. *J. Mach. Learn. Res.* **2009**, *10*, 719–742.
38. Bagirov, A.M.; Taheri, S.; Ugon, J. Nonsmooth DC programming approach to the minimum sum-of-squares clustering problems. *Pattern Recognit.* **2016**, *53*, 12–24. [CrossRef]
39. Karmitsa, N.; Bagirov, A.M.; Taheri, S. New diagonal bundle method for clustering problems in large data sets. *Eur. J. Oper. Res.* **2017**, *263*, 367–379. [CrossRef]
40. Khalaf, W.; Astorino, A.; D'Alessandro, P.; Gaudioso, M. A DC optimization-based clustering technique for edge detection. *Optim. Lett.* **2017**, *11*, 627–640. [CrossRef]
41. Le Thi, H.; Pham Dinh, T. The DC (difference of convex functions) programming and DCA revisited with DC models of real world nonconvex optimization problems. *J. Glob. Optim.* **2005**, *133*, 23–46.
42. Astorino, A.; Miglionico, G. Optimizing sensor cover energy via DC programming. *Optim. Lett.* **2016**, *10*, 355–368. [CrossRef]

43. De Oliveira, W. Proximal bundle methods for nonsmooth DC programming. *J. Glob. Optim.* **2019**, *75*, 523–563. [CrossRef]
44. De Oliveira, W.; Tcheou, M.P. An inertial algorithm for DC programming. *Set-Valued Var. Anal.* **2019**, *27*, 895–919. [CrossRef]
45. Gaudioso, M.; Giallombardo, G.; Miglionico, G. Minimizing piecewise-concave functions over polytopes. *Math. Oper. Res.* **2018**, *43*, 580–597. [CrossRef]
46. Joki, K.; Bagirov, A.M.; Karmitsa, N.; Mäkelä, M.M. A proximal bundle method for nonsmooth DC optimization utilizing nonconvex cutting planes. *J. Glob. Optim.* **2017**, *68*, 501–535. [CrossRef]
47. Joki, K.; Bagirov, A.M.; Karmitsa, N.; Mäkelä, M.M.; Taheri, S. Double bundle method for finding Clarke stationary points in nonsmooth DC programming. *Siam J. Optim.* **2018**, *28*, 1892–1919. [CrossRef]

 © 2019 by the authors. Licensee MDPI, Basel, Switzerland. This article is an open access article distributed under the terms and conditions of the Creative Commons Attribution (CC BY) license (http://creativecommons.org/licenses/by/4.0/).

MDPI
St. Alban-Anlage 66
4052 Basel
Switzerland
Tel. +41 61 683 77 34
Fax +41 61 302 89 18
www.mdpi.com

Algorithms Editorial Office
E-mail: algorithms@mdpi.com
www.mdpi.com/journal/algorithms

www.ingramcontent.com/pod-product-compliance
Lightning Source LLC
LaVergne TN
LVHW070043120526
838202LV00101B/418